CCSH LIBRARY

3 4199 00261 2170

Maxwell, Steve.
The future of water

MAR 0 8 2012

DISCARD

D1010704

the
Future
of
Water

A STARTLING LOOK AHEAD

by STEVE MAXWELL
with SCOTT YATES

Foreword by BRUCE BABBITT

The Future of Water: A Startling Look Ahead

Copyright © 2011 American Water Works Association

All rights reserved. No part of this publication may be reproduced or transmitted in any form or by any means, electronic or mechanical, including photocopy, recording, or any information or retrieval system, except in the form of brief excerpts or quotations for review purposes, without the written permission of the publisher.

The authors, contributors, editors, and publisher do not assume responsibility for the validity of the content or any consequences of its use. In no event will AWWA be liable for direct, indirect, special, incidental, or consequential damages arising out of the use of information presented in this book. In particular, AWWA will not be responsible for any costs, including, but not limited to, those incurred as a result of lost revenue. In no event shall AWWA's liability exceed the amount paid for the purchase of this book.

Project Manager: Scott Millard
Publications Manager: Gay Porter De Nileon
Technical Editor: Martha Ripley Gray
Production Editor: Cheryl Armstrong
Cover Art: Maurizio Marotta

Library of Congress Cataloging-in-Publication Data
Maxwell, Steve.
  The future of water: a startling look ahead/by Steve Maxwell; with Scott Yates; foreword by Bruce Babbitt.
    p. cm.
  Includes bibliographical references and index.
  ISBN 978-1-58321-809-9
  1. Water resources development. 2. Sustainable development. 3. Climatic changes--Environmental aspects. I. Yates, Scott. II. American Water Works Association. III. Title.
  HD1691.M38 2011
  333.91--dc22

                                            2011005552

MIX
Paper from
responsible sources
FSC® C014174

Printed in the United States of America

**American Water Works
Association**

6666 West Quincy Avenue
Denver, CO 80235-3098
303.794.7711 · www.awwa.org

To my children, Nick and Beck,
with the hope that their generation will be
more water-smart than preceding ones.

CONTENTS

FOREWORD

I became governor of Arizona in 1978 in the midst of a rainstorm. The rain continued for a week, then several weeks more. The Salt River, a desiccated scar of rock and gravel that cuts through Phoenix, suddenly came to life, topping its banks, tearing out bridges and crossings, cutting the city in two.

Today, thirty some years later, those floods are a distant memory. Throughout the Southwest, rivers are running low and reservoirs are drying up. Archaeologists deciphering tree rings from ancient ruins, and climate scientists modeling the onset of global warming, are warning that we are entering a severe and lengthy drought cycle.

I thought of these changes during my lifetime as I read the unsettling scenario in the Prologue of this book. Are we really entering a dystopian *Blade Runner* future of disappearing rivers and disintegrating cities? Or can we just continue to go about business as usual? My own Arizona experience suggests an unpredictable future, with surprises awaiting us out there.

Moving beyond the opening scenario, Steve Maxwell takes us straight into the realities of the water crisis that is now spreading through all parts of the country, and indeed of the entire world. Throughout the last century, we have populated the land without regard to natural limitations of climate and water availability, expecting there will always be another project to divert more water from somewhere else. In the classic movie *Chinatown*, the villain, Noah Cross, proclaims "either you bring the water to L.A., or you bring L.A. to the water."

Now, at the beginning of another century, we have about reached the limits of this transfer-at-any-cost policy. Schemes have been advanced to divert the Mississippi River to west Texas, to utilize the Great Lakes as a reservoir to irrigate Nebraska, and to channel waters from the Yukon River across the West clear down to the Mexican border. In these deranged dreams, we have at last come face to face with limits imposed by economic reality and ecological damage.

Among the many important themes of this book is Maxwell's clear and compelling discussion of the demand side of the water equation, that is, changing and improving the way we use water. The use of technology, not to build pharaonic supply-side projects but rather to drive efficiency, is among the most neglected of topics, and the author provides many insights into how we can moderate demand and deliver, use, and reuse water in new and innovative ways.

We use, by some estimates, 2,000 gallons of water to produce a pound of beef and 25 gallons of water to produce a can of Coke. We design expensive water utility systems to deliver drinking-quality water, 80 percent of which is used for watering lawns and waste disposal. And we have barely begun to explore agricultural water use efficiencies.

How can we bring about necessary changes in water use? The best place to begin is with simple economics, driving conservation and efficiency through pricing water at its full cost. In most parts of the country, the true costs of providing water are hidden in public subsidies that encourage excessive use and inhibit technological innovation. Throughout this book, the author takes us back to the importance of pricing water at its true cost.

For all the water management issues that we face, none are insurmountable. Water, unlike oil and other fossil fuels, is not a diminishing resource. The amount of water on this planet is fixed, and it is recycled in a continuous process from sky to land to ocean and back again into the atmosphere, to start the cycle again. However, the time is at hand to manage and use this resource properly. Maxwell's book is an excellent, indeed indispensable, guide to a sustainable water future, and it deserves a wide readership.

Bruce Babbitt
Former United States Secretary of the Interior
Former Governor of the State of Arizona
Washington, D.C.
December 22, 2010

❧

PROLOGUE

*I*t was late October, in the year 2111. As the early sun streamed into his bedroom window, Joe awakened to another hot, sweaty morning, in one of the still populated areas of northwest Los Angeles. As his wife rolled over and threw off the sheet, Joe lay in bed for a few minutes, reflecting on the events of the past few weeks before dragging himself to the water room. Today he was going to find out if he was one of the lucky ones or if he was one of the guys who would be laid off, as his company continued to cut back its West Coast water trading presence.

They didn't have much reason to stay in L.A. these days anyway, thought Joe. The last few family members that he and his wife Ellen had in the area had recently picked up and moved back to the booming Cleveland–Buffalo corridor, where a good bit of US manufacturing was now concentrated. And many of their other long-term friends had long since migrated back to the Midwest or found jobs with the North American government in the Winnipeg area.

In the water room, Joe flipped on the solar fan, then relieved himself into the male urine recycler. He heard the ultrasonic distillation tank kick in behind the wall—the low sound that, much to her consternation, always woke up Ellen. Joe listened as the tank disgorged a pint or so of clean water into the column above the sink. A few seconds later, he could hear the concentrated by-product stream discharge through the filter and into the crystallization reactor in the basement.

Joe rubbed his eyes and tried to remember what day it was. Tuesday—his day for a shower. He jumped into the moisture compartment, sealed the door and hit the "wet" button so that he could lather up. A couple minutes later, he hit the "rinse" button and luxuriated in a full 30-second stream of nice hot water (it had been sunny all day yesterday, and the water in the rooftop tank was still nice and warm).

A small electric pump kicked in after the rinse cycle and sucked the water through a floor pipe and over to the flush tank on top of the female urine recycling system on the other side of the bathroom. "OK, honey, you can get up now," said Joe as he stepped out of the compartment and started to dry off with his nanosorb wipes. Then he threw on some shorts and went into the kitchen, pulled out a bottle of fresh REWater, and mixed up some milk for his morning cereal. "Non-Fossil–100% Recycled–Save Our Continent" said the label on the side of the bottle.

Joe knew that it was probably going to be trouble at the office today. He'd been wrestling back and forth in his mind for several weeks about what to do when the time came. As a water trader, he knew the situation was only getting worse, and there was no longer much sense in pretending that much of a market remained in the L.A. area. There just wasn't that much trading activity going on anymore.

Joe chewed on his cereal. It was only about two hundred years ago, he thought, that this whole place was just a big pile of sand, with maybe a few cypress trees. And it may not be too long before it's just a pile of sand again. We came in and dumped some water on it, and almost overnight it became one of the most heavily populated and concentrated centers of economic power anywhere in the world, or for that matter, in the history of the world. But those days are over. That was sort of a mirage.

Now, Joe worried, Southern California seemed to be dying almost as quickly as it had been built. An armed insurrection in Colorado in the late 2050s had eventually led to the landmark Supreme Court case in 2066 that cut California's share of the declining Colorado River flows, and soon thereafter the massive Southern California economy began to decline. And after the protracted drought and the huge population losses of the 2080s, many of the area cities decided they could no longer afford to provide the public with water, and they shut down many of the water distribution systems in the Inland Empire. Thankfully, Joe's home was in the Hollywood Hills area—one of the few that still had a centralized water distribution system of sorts, albeit in a state of ominous disrepair.

Extensive water pipelines from the northwestern United States and Canada were built during the middle years of the century to try to maintain irrigation in the Central Valley and to support other population centers in the American Southwest. However, as problems worsened, Americans had finally recognized what skeptics had been saying for decades: the Southwestern oasis had been a mirage from day one.

With rising seas, Los Angeles experienced wide-scale seawater intrusion into its key coastal freshwater aquifers, exacerbating the surface water shortages that were already a tightening noose around the city's neck. That pretty much killed the coastal communities. Then the ill-fated effort to take down the Hoover Dam in the 2080s caused vast and unforeseen ecological impacts downstream and further depressed the economy—and perhaps more importantly, the general mood—of Los Angeles. It was sort of like a historical swing of the pendulum, thought Joe—but opposite to the way it had swung a hundred and fifty years earlier, when the new dam buoyed the development of Southern California, and an American victory in World War II brought the United States out of the Great Depression.

By the late 2090s, the rich people were beginning to have water trucked into their neighborhoods from the north. As the oasis of beautiful gardens and golf courses dried out and decayed, the basin was once more gradually turning back into a desert. Joe remembered once when he had been able to fly into L.A. from a meeting back east; he had looked down as they came in above the virtual ghost town areas of eastern San Bernardino and Riverside counties. From the air, Los Angeles was starting to resemble a mummified skeleton; the freeways looked like aging, brittle bones pushing out against a stretched and drying skin. As more people left for Vancouver, Ketchikan, Winnipeg, and beyond, some of their abandoned homes were being taken over by squatters arriving from farther south in Mexico and Central America.

The once-blooming desert cities of Phoenix, Tucson, and Las Vegas also fell into decay quite quickly; once the major reservoirs began to decline, and a couple of the key pipelines were shut down, they couldn't hope to survive. Other smaller southwestern cities dependent on external water sources were turning into ghost towns by the end of the century—reminiscent of the time, two centuries earlier, when the gold and silver veins ran dry. After the nationalization of the energy industry, the Southwest became even less habitable; a ban on air-conditioning and widespread water shortages essentially left no alternative to northward migration.

As the climate slowly dried across the southern United States, agricultural fertility—and later, large segments of the human population—started slowly but surely to migrate to the North. As the winters shortened, the forests of the American West were decimated, first by pine beetles and later by massive wildland fires that attacked the weakened forests. In fact, the Western skies were so darkened with smoke for many years late in the century that some scientists even predicted a long-term reversal in the warming trend.

Winter snowpack declined, and with that many of the Western cities saw their primary source of water decline. Year after year, the streams ran dry by midsummer, reservoirs dipped, and fires overran the ecosystem. In earlier days, glacial melt had helped supplement the river flow in dry years, but now the glaciers were mostly gone as well.

In some regions sufficient water still ran, but reservoirs couldn't be built fast enough to store the spring runoff. Water storage capacity started to become the key driver behind commerce and demographics. Most of the Rocky Mountain area population was beginning to migrate toward the Midwest or into Canada by the end of the century for just this reason. The net result was that experts were predicting that by the year 2120, the demographic center of the North American continent would be located somewhere just southeast of the capital of Winnipeg—a huge shift in bodies to the north.

The major financial institutions were among the first to truly understand and recognize the eventual economic impact of the warming climate and the rising seas. Several of them helped to underwrite the forty-niners— the firms that led the great "Ice Rush" of Greenland in 2049, two hundred years after the gold rush. The big financial firms began to withdraw from New York, with several firms moving their people to Cleveland in 2061. Joe's own employer moved to Buffalo a couple of years later. The economic power and cultural dominance of the major Eastern cities began to decline, while what had long been called the "rust belt" cities enjoyed a measured revitalization.

Joe was jolted from his thoughts when he heard the female recycling system clank several times and discharge a quick shot of water onto the tomato plants on the south side of the house, outside the kitchen window. Ellen emerged from the bedroom dabbing her face with a damp nanosorb that she had retrieved from the walls of the water compartment. "I don't know if I can stand another day of these temperatures," she announced. "It's almost November, for God's sake." "It's supposed to cool down by the weekend," Joe replied.

It wasn't much of a surprise, mused Joe, as he downed his last few sips of coffee and started to dress for work, when the United States finally "invaded" Canada in 2083. Actually, it wasn't so much an invasion as a friendly takeover. The president of the United States at the time, Jon-Michael Simpson, was from northern Vermont and of partly French-Canadian ancestry. He'd called his Canadian counterpart to tell him of the impending action, and then made a rare digital holographic address to the nation prior to sending troops over the border on a warm Sunday evening.

Despite a few isolated pockets of token resistance in the Great Plains region, the American soldiers quickly took command of the key water and energy resources in Ontario, Alberta, and British Columbia. Once these critical resources were secured, the United States and an agreeable Canadian prime minister sat down at a negotiating table in Minneapolis and decided that—acting together with the combined military and economic resources of the United States and the natural resources of Canada—they would form the dominant political entity in the new world order.

It later turned out that Canadian premier Roland Garth, weeks earlier, had essentially asked the United States to invade his country—to do what many practical people in both countries realized should be done in order for them both to better survive in the rapidly changing geopolitical balance. For the Canadian nationalists, this came to be considered the greatest act of political treachery since 1941, when President Roosevelt reportedly ignored clear warnings of an impending attack and let Pearl Harbor be decimated in order to build public support and draw the United States into World War II.

Compared to other more fragile water- and energy-based coalitions that were emerging in other corners of the world, the Canadians and the Americans meshed quite well. Between them, they had the wealth of water and natural resources and the ingenuity and drive to build a commanding political presence. For decades, the North American population had been gradually migrating north and toward the center of the continent. The migrants sensed that a central locus might provide them some level of protection in the future and looked over their shoulders: Winnipeg was declared the capital in 2091. Winnipeg could probably be a pretty livable place if we had to move there, thought Joe. It still got a little cold in the middle of the winter, but it was pleasant most of the year, and was still a comfortable thousand miles or so from the deserts encroaching from the south.

The Canadian water and energy resources were extensive, enough to support the continent's population for a while, but the North American long-range planners were beginning to realize that they would eventually need more. With that in mind, they had begun, actually many years before, to eye Central Africa—the world's poorest and economically most dysfunctional area. The northern half of Africa was basically unpopulated by the beginning of the twenty-second century. Many of its original people were gone; the more resourceful fled the continent entirely, while most of those from the more impoverished countries perished during the interminable trek southward. The southern tip of Africa also dried and then politically disintegrated during the latter half of the century, with the wealthier people moving north. Now, the steamy and formerly backward center of the continent held out an allure as one of the world's richest sources of water and minerals—and many predicted it would become a key economic and financial center of the future.

Kinshasa as a key economic center of the world—who would have guessed that a hundred years ago, Joe asked himself. And all just because it happens to have a lot of water. Nowadays, Joe and Ellen had gotten comfortable using about 20 gallons of water per day. Joe's grandfather, on the other hand, even in the drought of the 2030s, had used close to 200 gallons a day. But no matter how much they used, water was still expensive; they were spending a lot more on water these days than on food. If water climbs above $7 a gallon, Joe thought, we can't afford to stay here—even if I do manage to keep my job for a while longer. We're already spending almost $150 a day on it. We'll simply have to follow the crowd and move to where the jobs are—where the water is.

And Ellen was very worried about another issue that was starting to hit the news around L.A. For decades, scientists had been routinely testing the local surface waters, and they were finding higher and higher concentrations of various man-made pharmaceutical compounds and a multitude of industrial chemicals. Sure, the water only contained minuscule amounts—one part per trillion, or so—but modern instruments were finding this stuff in waterways all over the state, even up in the mountains. It appeared that people had unknowingly ingested these materials for years, if not decades, and the scientists were concerned that it was having a grave impact on human reproductive and endocrine systems. Everyone knew that sperm counts among young males were falling dramatically—and now the experts were blaming these mystery water contaminants for the sharp drop in birth rates in the United States.

No one really understands what the hell is going on here, thought Joe, but something is definitely happening. Over time, mankind was clearly becoming less fertile. That's why the Continental Government had recently mandated separate male and female urine recyclers. Apparently, the female body was better at processing and withstanding these tiny contaminants than the male body was, and the government—in all of its wisdom—determined that the concentrated male by-product stream should not be used in the direct recycling system or to irrigate plants for human consumption. Hence the crystallizer in the basement. If current trends continued, said the scientists, human fertility may be doomed. Maybe all of this was just God's way of solving all the problems we've created for ourselves, Joe thought wryly.

Why did humans ever decide they wanted to live in this place anyway, Joe wondered. All you've got to do is just look outside and you can see that it's going to turn back into one big damn desert. It was a pretty inhospitable desert when those first folks showed up, too. And now they're actually talking about turning off the Vancouver pipeline. Joe and Ellen both knew there was no way they could afford to have their water trucked in, like the remaining wealthy families down in Hancock Park were doing.

He looked out the window at the gravel in the yard and the small green patch of artificial turf where he and the kids wrestled on the ground and putted a golf ball around. The cactus plants that he and Ellen had planted almost a year ago looked wilted. We'll probably have to dig those up and toss them before long, he thought. Not enough water. Maybe those idiots who are talking about opening up a water resort on the moon aren't as crazy as everybody thinks they are, Joe muttered under his breath, as he got up to head to the office. Maybe I should see if they're hiring.

OK, maybe not the moon, but Joe was increasingly thinking about following a couple of his buddies, and moving to northern Canada, or maybe even to Africa. There were good water jobs in both places. That's where the water is; that's where the opportunities are, his friends had said. That's where the future is. One of these days, I really am going to have to sit down and talk with Ellen about all of this. She wants to get the kids out of here, *anyway. But not today.* "Bye, honey, I'll see you tonight," Joe said, as he slipped out the door and into the heat.

℘

Does this scenario sound a little crazy and a little scary? Scary—certainly. Crazy—maybe not.

Hopefully, our future will not turn into this kind of dystopia. Maybe it will turn out that with smarter water conservation systems and with wiser water management policies, the California lifestyle that we know today can remain vibrant. Maybe in a hundred years, California's economy will be driven by a vast array of solar panels in the Mojave Desert. Perhaps the Central Valley of California will be returned to its original state, and people will buy most of their food from the Midwest, acknowledging that rain-fed agriculture is more sustainable. Maybe millions of rooftop harvesting systems will collect rainwaters and direct them through the home's internal uses and then onto extensive vegetable gardens in the front yard. Perhaps a hundred years from now, desalination powered by the waves will help provide enough freshwater to keep homes and industry functioning just fine. Perhaps we'll have genetically modified grass that can be irrigated with seawater, so the children of the future can play on grass, not plastic. Maybe storm waters will be collected and delivered to distributed neighborhood water treatment plants, rather than running unused to the sea. Perhaps the word *waste* itself will fade from our vocabulary, and we will see in everything a resource—a possibility. Perhaps.

As we will see in the following pages, the availability and the price of water will increasingly come to dominate economic, political, and social trends in the future. Unlike other commodities, water is infinitely renewable, but its supply is essentially fixed, and we have no substitutes whatsoever for the critical role that water plays in each of our lives. And no matter how many people end up living on this small planet, we are always going to have exactly the same amount of water.

℘

8

CHAPTER 1

LOOKING BACK ON THE FUTURE OF WATER

It's obviously difficult to look forward a hundred years and guess what life might be like. But looking back a hundred years may give us a few insights. If we transport ourselves back to 1910, we may get a better understanding of just how hard it would have been to predict what our world looks like today.

LOOKING BACK TO 1910

In 1910, ninety-two million people lived in the United States; in 1810, only seven million people had lived in this country. Today, almost three hundred ten million people live in the United States. But we still have the same amount of water that we had in 1910—the same amount we had in 1810.

Today, about the same number of people live in the desert city of Las Vegas as lived in the water-rich city of Chicago in 1910. Las Vegas scarcely existed in 1910, but today the area has two million people. In 1910, Phoenix had fewer than ten thousand people, and Arizona wasn't even a state; today, the area is home to more than four million people. Los Angeles had three hundred twenty thousand people in 1910—today, the L.A. area is home to about twenty million people.

But we have the same amount of water in the United States now as we had in 1910.

In 1910, electricity was just beginning to find its way into the home; the average US citizen used 271 kilowatt-hours of electricity per year. By the end of the century, the average US citizen used about 13,000 kilowatt-hours per year. Today, the state of California uses over 15,000 gigawatt-hours (that's 15 billion kilowatt-hours) of electricity just to treat water and pump it around the state. Most of that water goes to the booming and increasingly thirsty cities and to the agricultural operations in the Central Valley.

A hundred years ago, the Central Valley of California was primarily known for being flat and barren, with stifling heat in the summer and six months with little or no rain. On less than 1 percent of the total farmland in the United States, that valley today produces almost 10 percent of all the agricultural products sold in the United States. The value of total US agricultural output in 1910 was $7.7 billion; by 2009, it had grown to $331 billion.

In 1910, only a few years had passed since the Wright brothers became known for something besides their bicycle shop. Sixty years later, men were walking on the moon. A hundred years later, in 2009, two of NASA's biggest accomplishments were the installation of a wastewater recycler on the International Space Station and the discovery of water on the moon—something NASA said would make travel to Mars and beyond much more likely.

Only a few people had radios or telephones in 1910. The refrigerator hadn't yet been invented. Automobiles were the province of the rich and famous. In 1910, there were half a million automobiles in the United States; today, over thirty million vehicles are registered in California alone. In 1910, the average US laborer earned $15 a week. The United States had just surpassed a total of a thousand miles of concrete highways. In 1910, the Chicago Cubs lost the World Series and Harvard College won the NCAA football championship. In 1910, it would only be ten years until women were given the right to vote. A long time ago, indeed.

So, back in 1910, it would have been pretty tough to guess what 2010 might look like. From a water perspective, in 1910, most people in the world lived close to or downhill from relatively clean and abundant water sources. People didn't worry much about water because it was still relatively abundant and clean, and for most folks it was essentially free. Most of their food came from crops that were fed by the rain. Most people lived on farms or in rural areas.

In 1920, Yale professor Frederick Haynes Newell published a book entitled *Water Resources, Present and Future Uses,* in which he wrote extensively about water conservation.[1] But in 1920, *conservation* meant something quite different. To Newell, conservation actually meant usage and exploitation of the resource; that is, not wasting the water by letting it run downstream unused. For instance, in one place he writes, "Floods [were] restrained by the Roosevelt Reservoir, Arizona, water otherwise destructive held in part for future use in generation of electric power, for irrigation of arid lands, illustrating double or triple benefits of conservation." He doesn't mention environmental issues, in-stream flow levels, or biodiversity impacts. In fact, in the widely held view at that time, major dam construction projects were the best way to pay homage to the concept of water conservation.

Or consider early predictions about the Ogallala Aquifer, located under the Great Plains, which we will discuss in more detail later. After the economic and social disaster of the Dust Bowl, government officials and insurance agencies were anxious to make sure history didn't repeat itself. They waxed optimistic about the future of the Great Plains and the abundant groundwater underlying the center of the country that would irrigate the land forever. In a report dating from as late as 1959, Travelers Insurance Company said that pumping up water from the Ogallala would be sufficient for farmers to irrigate their crops, hypothetically forever. That report made no mention of the fact that water was being removed much faster than it was being replenished; nor was there any thought that this practice could, at some point, cause the aquifer to run dry. It was well into the 1980s before officials began to worry about the drawdown effects of massive agricultural withdrawals on the Ogallala. And as we will see, even today the unsustainable mining of this groundwater continues largely unabated. Only as the water levels sink so low that it becomes prohibitively costly to pump up the water will we see this rate of extraction drop.

On September 30, 1935, President Franklin Delano Roosevelt gave a stirring dedication speech for the opening of the Hoover Dam, then known as the Boulder Dam. He said, "What has been accomplished on the Colorado in working out such a scheme of distribution is inspiring to the whole country. [The federal government has] ... constructed a system ... which will insure to the millions of people who now dwell in this basin, and the millions of others who will come to dwell here in future generations, a just, safe, and *permanent* system of water rights" (italics added). He continued, "The mighty waters of the Colorado were running unused to the sea. Today we translate them into a great national possession."[2]

LOOKING FORWARD

Seventy-five years later, water levels have fallen dramatically, and the Lake Mead reservoir behind the Hoover Dam is now less than half full. Sediments continue to build up on its bottom, while drought and evaporation have depressed its surface. A "bathtub ring" of bleached white rock extends more than a hundred feet above the current surface of the reservoir, showing where the surface of the water used to be. In October 2010, the lake hit its lowest level since the reservoir was originally filled in the 1930s. Officials are hurriedly installing a new and lower water intake, at a cost of almost $500 million, so that the city of Las Vegas will be able to continue to draw its water from the lake. Officials are worried that if the reservoir drops much lower, Hoover Dam may have to scale back its generation of electricity, with a dramatic impact on energy prices and availability throughout the Southwest. A senior policy analyst for the Natural Resources Defense Council commented, "This is the place where the mega-dam began, and it may be the place where it ends."[3] Times change.

Attitudes about water and public awareness of water challenges have changed dramatically in the past few decades and are in a rapid state of flux today. At best, we can only make some educated guesses about what the next hundred years may bring in terms of water uses, water challenges, and water prices. For example, right now it may only make sense to recycle urine into drinking water in an orbiting space station, but in a hundred years most toilets may work the same way. Right now, most crops are irrigated with sprinklers, and drip systems are just becoming popular. Both may seem like relics in 2100; in the future, we may be watering all our plants from the roots up, with underground drip systems.

Right now, water from the ocean is only rarely used, after being treated at high cost, for drinking or to water crops—but will we increasingly rely on desalted water in the future? Or will we develop crops that can be irrigated with seawater, on a scale sufficient to feed the massive populations of countries like India and China? Right now, the average apple eaten in the United States has traveled 1,550 miles, and an apple is 84 percent water. As energy costs continue to inexorably rise, will people instead be growing their own apple trees and irrigating them with rain captured from their own roofs? Or eating fewer apples? The questions go on and on.

But we can say one thing without much doubt: water will be a much more critical determinant of that future than it has been in the past. The entire sweep of Earth's history is inextricably linked to water. The first life on Earth formed in shallow tidal pools at the edge of the primordial seas. The evolution of life on Earth followed the ebbs and flows of water. Human history closely reflects the natural system of waterways that we inherited—water that we used for transportation, power, defense, and sustenance. Throughout most of our history, our major cities have always been coastal ports or population centers that sprang up at the confluence of major inland rivers.

Indeed, it is only within the past hundred years or so that we have really been able to begin to sever our civilization's historical marriage to natural water systems—our need to live close to water. As new technologies have made possible the wide-scale transfer and movement of water, the human race has been able to branch out and relocate itself, to live and work far away from natural sources of water. Rural electrification and modern electric pumps have enabled humankind to turn arid deserts into productive farmlands. New engineering techniques have allowed the construction of massive structures like the Hoover Dam, to provide power, flood control, and—most importantly—water where none existed before. Huge cities have sprung up in the high arid deserts of the Western United States. With these technological advances and economic changes, our world has been transformed.

But now we are beginning to experience the challenges that this separation of humans from water really represents in terms of unwise use of resources, and we're seeing that perhaps this divorce was not destined to last very long. We are now beginning to reach the limits of what all these pumps, dams, reservoirs, and irrigation systems can do for us. We are beginning to understand that perhaps it doesn't really make any long-term sense for our major migration trends to be away from water-abundant regions and into deserts. We are beginning to understand the fallacies of trying to grow rice in some of the most naturally arid regions of the world, like Australia did until recently. We are beginning to realize the water isn't free, and we are just beginning to realize that it may make sense to take water into account when we make bigger economic, political, and social decisions. In short, it may be time to restore our historical relationship, to repair our marriage to the natural systems of water.

Water issues are generally local issues, and the intensity or immediacy of water problems varies widely from one region to another around the country and around the world. Some regions have catastrophic water problems staring them in the eye, while other areas are blithely unaware of any challenges. However, paraphrasing from the old bumper sticker, while we need to act locally, we need to think globally. And from a global perspective, there is little doubt that we are now emerging out of a historical age of water *development*, and into a new era of water *allocation*. Said another way, we've already developed and exploited pretty much all the water we have, and now it's time to figure out how to use it more wisely.

FOUR MAJOR AND RECURRING THEMES

This book is aimed at answering some of these questions about the future of water. We have tried not to take too much of a Chicken Little approach here; even though we face extremely serious and daunting challenges in the future, we have attempted to illustrate not only the problems but also potential solutions or ways of meeting the challenges. As you read through the book, you'll see that we've organized our thoughts and analysis around four major and recurring themes.

First, we will see that water is increasingly and rightfully becoming recognized as one of the key criteria, or factors of production, in industrial manufacturing, public policymaking, and personal decision-making. Economists have traditionally thought of labor, capital, and energy as the critical inputs to economic decision-making and our rising standards of living; we will soon see water considerations join this club as well. And, as we'll see, water imperatives will often conflict with energy or capital imperatives. A sounder and wiser water future will imply balancing many water issues and challenges against similar capital, labor, and energy challenges.

Second, we'll see much more emphasis given to the concept of the *water footprint*, or total water impact of everything we do and everything we own or consume. Only by understanding the *full* water impact of what we do—the total amount of water that goes into a product or that we utilize in a given behavior—can we move toward wiser decisions and more efficiently allocate our scarce water resources. The total amount of water that it takes over the full life

Looking Forward

- *Water will increasingly be viewed as a true factor of production—like energy, labor, or capital—in manufacturing, policymaking, and economic decision-making.*
- *Water usage will be evaluated in a more holistic manner, from the perspectives of virtual water and the water footprint.*
- *Boundaries between different types of water—drinking water, wastewater, rainwater, storm water, source water, groundwater, seawater—will fade, with the emerging management of one water.*
- *The price of water will rise to reflect its true cost and value.*

cycle to produce a given product or service is increasingly referred to as its *virtual water* content.

Third, although we will talk about lots of different kinds of water in this book—drinking water, wastewater, rainwater, storm water, source water, groundwater, seawater, contaminated water, and so on—in reality, all water is just water. There is truly only one water. We'll see more and more recognition of this in the future, and we'll see a gradual crumbling of the historical boundaries between, for example, drinking water and wastewater. Before we can solve our myriad challenges, we need to begin to think more holistically about water.

Finally, a major theme will revolve around the price we pay for water, a figure that will likely increase rapidly and that will dramatically change our priorities and our behavior in the future. Today, most of us pay a price that does not really reflect the true costs of delivering that water and certainly not the true value of that water to us. If we continue to assume that water is free, or almost free, we will tend to waste it and not pay much attention to how we use or conserve it. Once water prices rise high enough to affect our wallets, our attitudes and behavior will start to change, and we will be forced to become better stewards of this scarce resource.

As the saying goes, "It's hard to make predictions, especially about the future." In this book, we'll nevertheless give it a try.

ℰℐ

[1] Frederick H. Newell, *Water Resources, Present and Future Uses*. New Haven: Yale University Press, 1920.

[2] Franklin D. Roosevelt's Dedication Day Speech, September 30, 1935. Available at University of Virginia American Studies Program and the 1930s Project. http://xroads.virginia.edu/~MA98/haven/hoover/fdr.html.

[3] Barry Nelson quoted in Felicity Barringer, "Lake Mead Hits Record Low Level." *New York Times*, October 18, 2010. http://green.blogs.nytimes.com/2010/10/18/lake-mead-hits-record-low-level/.

CHAPTER 2

THE FUTURE OF WATER USE OUTSIDE THE HOME

Most of us consume most of our water in and around our homes. So that seems like a good place to start, in terms of our exploration of the future of water. We'll talk about lawns and gardens, and in the next chapter, we'll discuss newfangled washing machines that don't use water, as well as clothes that never need to be washed, the relative environmental benefits of driving a Hummer, and the disaster of nonrecycled toilet paper. First, let's talk about water usage around the outside of our homes.

LAWNS: THE AMERICAN DREAM

For horticulturists and plant biologists, the very idea of a lawn is just plain strange. For essentially every other plant that biologists study, the full life cycle of the plant is explored and sometimes managed so that the bounty of the plant is fully realized. For instance, three cousins of turf grass are corn, alfalfa, and wheat. Those three plants make up a huge percentage of farmland in the United States, and to varying degrees they are the big three planted around the world. Biologists have extensively studied how to most efficiently support and grow those plants through their entire life cycle. Especially with corn and wheat, the goal is to help the plants reach maturity and produce seeds, for it is the seeds that we value the most for use in food for humans and animals.

With turf grass, however, it's a different story, and the whole process is turned upside down. Experts in growing and tending the lawns for the relatively wealthy people of the world focus instead on how to discourage the plant from seeding. The biological equivalent is like something out of science fiction: How to grow an organism that has a natural life cycle, but don't let it live out that life cycle. Keep the organism perpetually in a stage of prepuberty.

It's possible to do this, of course, because turf grass can live, grow, and multiply through its roots. It can reproduce without seeding—and in lawns, it's never allowed to seed. Most of the varieties of grass used in lawns these days have been developed by scientists, many of them working for the US Department of Agriculture, to spread through their roots.

To continue to stay alive in this constant state of prepuberty takes an extraordinary effort on three fronts. First is the management of the plants by constant cutting before the grass can begin to seed. Next, usually, is the application of various manufactured chemical fertilizers, pesticides, and herbicides that nurture the plant in its perpetual juvenile state, keeping it looking lush and verdant no matter what the local climate.

The third front, of course, is water, and lots of it. All plants require water, but few are as thirsty as turf grass. The water must soak into the roots to keep the grass growing, but the blades that are so popular above the earth do a pretty good job of keeping much of that water from reaching the roots. More than half the water applied to a typical lawn during a hot day will be lost to evaporation; indeed, much of the water sprayed onto a lawn overnight will also end up evaporating during the day. Besides the visual beauty, this is an important part of why grass is so popular: it just feels physically cooler to spend time on the top of it. Walking on or near grass feels cooler than the surrounding sidewalks or streets because the grass itself is evaporating water. When tiny microdroplets of water evaporate, they essentially take some heat with them. That is to say, walking on a lawn doesn't just feel cooler; it actually is cooler, because of the evaporating water.

Water utilities can detect with some precision how much water is used on lawns. It's easy to track, especially those in areas where it's too cold to irrigate in the winter. The amounts of water used inside the home in January and in July are essentially the same; the difference between the January water bill and the July water bill usually reflects the amount of outdoor watering. And nearly all outside water use is for lawns in the United States—essentially the only country in the world where lush and large lawns can be found in every town, village, suburb, and city.

The amount varies, but in general, most American families use nearly double the amount of water outside the home as inside—and it's mostly to keep their grass green. Water also goes to wash cars, fill swimming pools, and so on, but by far the biggest use of water is to irrigate turf grass. Interestingly, water features like fountains, ponds, and artificial waterfalls use less water than most lawns because they are filled once and don't need much to refill what's lost to evaporation. Flower and vegetable gardens typically require less than lawns even though they are producing flowers and food. Trees rarely require watering beyond what they get from their extensive root system, though to be sure they also get the benefit of some of the water that runs down through the lawn unused.

So for most of us, it's the green, green grass around our homes that consumes the biggest chunk of the water we use. In that light, the question, "What will lawns look like in the future?" is worthy of our attention.

Well, soak up a good long look at your lawn today, because the future doesn't look bright for wide expanses of green grass surrounding every home. The future is not always clear, but this is one area where the future is fairly easy to predict. The costs of finding, treating, and delivering water are going up rapidly and will continue inexorably to increase in the future. We will always need water for drinking, cooking, and cleaning. Hence, in the future, we will need to triage or set priorities and start cutting out some less essential uses of water. No matter how much we may like our big, green lawns, we just won't be able to afford them in the future, at least not as big as they are now.

Indeed, money is what's really at the root of why lawns have become so popular in the first place, especially in the United States. The lawn originated centuries ago with the British and the French, but only on the estates of the very rich. Societal strictures made those estates impossibly out of reach in the Old World. Our founding fathers envisioned lawns spreading throughout the New World, but with an egalitarian flair. The lawn has been an ingrained part of the American experience since the earliest days. George Washington had a wide lawn at Mount Vernon, and—since motorized lawnmowers weren't invented yet—he encouraged deer to come and graze to keep the grass from growing too high. Thomas Jefferson laid out the plan for the University of Virginia himself and wanted all the buildings to surround what is known to this day by the name he gave the wide strip of grass: The Lawn.

Ecologist John Falk, working for the Smithsonian Institution in the 1970s, developed a theory of what he called the Savannah Syndrome.[1] He suggests that as humans roamed Africa in search of food, they came to be comforted by vast plains of grass with occasional clumps of trees. With thousands of years of this kind of hunting and gathering, the desire for that kind of environment became imprinted in our DNA. That may or may not be, but even if so, it doesn't explain why the citizens of the United States, more than the residents of any other country, are so "addicted to grass."

Perhaps the first true suburb in America—Riverside, Ill., outside of Chicago—was designed in 1868 by Frederick Law Olmsted, famous for designing Central Park in New York City. Riverside featured wide lawns with no fences and curving roads. Every suburb since then has been more or less a duplicate of that one. Hand in hand with Olmsted's vision, according to Michael Pollan in his book *Second Nature: A Gardener's Education,* was Frank J. Scott, who wrote in 1870 that "a smooth, closely shaven surface of grass is by far the most essential element of beauty on the grounds of a suburban house."[2]

As popular as lawns were for the small number of suburbs, it was only after World War II that they really grew in size and popularity along with a burgeoning suburban population. The other phenomenon that took off during that period was that quintessentially suburban status symbol, the golf course. Only officers were used to playing golf after World War I, but by the time World War II had ended, enlisted men were taking up the game, too. Indeed, with the encouragement of the US government, ten golf courses were built just in Japan for the entertainment of the GIs who remained there after the war. When those same soldiers came home, they often settled around one of the biggest trends in suburban housing: the country club or golf course community. Some seventeen thousand golf courses now exist in the United States, according to *Golfweek* magazine, up from about four thousand in 1950.[3] Even those homes that weren't on or near an actual golf course were landscaped to look like they might have been, with growing expanses of lawn surrounding every new home.

Indeed, the lawn became a central part of the culture of postwar living in America. "The grass is always greener on the other side of the fence" became a cliché precisely because it was such a common experience. By 2005, analysis of satellite images by NASA showed that about 32 million acres of lawns were being watered and mown in the United States. That's about three times as much land as that which is irrigated to grow corn in the United States[4]—by far the largest irrigated crop, according to figures from the USDA.[5] The golf industry is a $76 billion business, according to the National Club Association, larger than the motion picture and music industries combined. Just the lawn care industry itself is now about $25 billion, according to industry estimates.[6]

TRENDS IN LANDSCAPING AND WATER SAVINGS

It's not clear just yet when we actually reached "peak lawn" when the total amount of acreage devoted to grass started going down, but the rate of growth has certainly slowed. The trend in new housing developments is for individual homes to have smaller and smaller lawns, with more emphasis placed on central parks and public areas with grass. Cumulatively, such park areas do not add up to nearly as much grass as in more traditional developments, where each home can have a half an acre or more of grass. Furthermore, when tended to by experts, the amount of water used on parklands is typically less than that used on home lawns because these experts have tools to understand the precise amounts of water needed. They can, for instance, apply water at more precisely timed intervals based on moisture sensors in the ground. Professionals also can use devices that tweak the amount of water applied based on recent rainfall, the maximum temperature of the day, and other factors. Finally, the economies of scale kick in much more quickly, so separate irrigation systems using household gray water or other sources of used or recycled water—as opposed to treated drinking water—can be used on the grass in larger public parks.

Expect those trends to continue as new neighborhoods are developed with smaller individual lawns and more emphasis on green spaces. The planted common areas in neighborhoods, in parks, or along boulevards in subdivisions will less often be blanketed in a sea of uniform, green grass. Instead, expect to see much of the world start to look more like Tucson, Ariz., with rocks, succulents, and drought-tolerant shrubs.

Common areas likely will be irrigated using technology that is radically different from what we have today. In Florida, for example, the city of Leesburg began a test in 2009 of a new system for irrigating common areas. Instead of using soil, ornamental species were planted in sand on top of a giant sheet of plastic. The plastic was contoured so that water drains to a central area that becomes a pond. Periodically, a series of pumps take water from the pond and distribute it directly to the roots of the plants in the sand. The plants drink what they need, and the rest drains back to the center of the pond. The entrepreneur and tree farmer behind this idea, Rufus Holloway, believes that as little as 20 inches of rainfall a year will be needed to recharge the system. That's well below the usual rainfall in Florida, even during drought years.[7]

There's no reason a system like that couldn't work in an individual family's yard. The up-front costs are high compared to just continuing to irrigate a lawn at current water rates, but as water rates start to climb, or if the homeowner is installing a new landscape anyway, the costs may not be that much higher. Such approaches could all but eliminate the need for treated-water irrigation in many parts of the world.

And for cities lying near an ocean, it may be possible in the future to grow grass with regular old seawater. Researchers around the world are working on this, but especially in Australia—in offices such as the Future Farm Industries Cooperative Research Centre based in Perth. Researchers there have developed grasses for food production that cross sea barley grass with wheat using cytogenetic techniques that change the plants at a cellular level, in essence allowing them to live on seawater.[8] Wheat is a cousin of the turf grasses, so it's not hard to imagine that genetically modified grass irrigated with seawater could be developed and planted in Jacksonville, Houston, Los Angeles, and other coastal cities with significant freshwater shortages predicted over the coming decades.

As water rates climb, not only will the cost of having a lawn increase for the typical family, but so will the cost of food. This means that another likely trend is something that's seen as slightly subversive or counterculture right now—the front-yard garden. Today, the sight of a working garden in the front of a home is almost jarring. In many cases, the homeowners association prevents such a garden. However, times are changing, and front-yard gardens may once again be common in twenty-five or fifty years. As food prices rise, more and more people are likely to realize that as long as they are pouring expensive water on their yards, they might as well get something tangible in return, beyond the psychologically pleasing expanse of green grass. (We'll discuss some of these trends in the next chapter.)

The suburbs of the United States are not going away. Neighborhoods typically change their makeup slowly once they are established, and typical American families will continue to occupy subdivisions created in the 1960s going forward. But the difference from a water perspective is how much of the land outside of those homes will be covered by grass.

This speaks to a caveat that we'll have to touch on various places in this book: obviously, the future is not going to look the same everywhere. In the past, we may have had more of a one-size-fits-all view of the world. If you had asked someone in 1960 what would surround nearly every home in fifty years, the answer would have been lawns, lawns, and more lawns—green grass surrounding every home, school, and business and covering all of every park and golf course. The golf industry has always been a leader in spreading the good news of grass. That's an answer that would have made perfect sense, and it has pretty much come true all across America.

CHANGING ATTITUDES

Although rainfall level, humidity, temperature, insect life, and soil itself are all radically different in, say, Seattle, San Diego, Pittsburgh, and Orlando, new housing developments and landscaping in all those cities look startlingly similar. Why? Well, part of the answer is a cultural norm that has been heavily promoted by the turf and golf industries. It has become so ingrained that it has even been made part of the code imposed on homeowners by homeowners associations and local governments. In many cases, associations *require* that homeowners keep a percentage of their property covered in grass. Stories of homeowners fighting city hall or homeowners associations switching to less water-intensive landscaping are becoming common all over the United States, especially in the West.

In the future, we will probably throw out standardized solutions in favor of local innovations that respond best to local problems. The other main factor at play here is energy—a topic that comes up over and over when we're talking about water, and that, as we'll see, is fundamentally interwoven with water issues. For a turf lawn to grow, water is certainly needed, but so is energy in several forms. For one, many of the fertilizers, pesticides, and herbicides that are put on lawns to keep them looking good are petrochemical derivatives, meaning that energy is effectively being diverted from other purposes to grow grass—energy that must be pumped out of the ground or from beneath the the ocean floor. Lawnmowers often have tailpipe emissions worse than most cars, and they often are driven around cities in trailers hooked to the back of gas-guzzling trucks. Green grass is hence not really all that "green." Or, as Michael Webber wrote in

Scientific American, "Someday, we might look back with a curious nostalgia at the days when profligate homeowners wastefully sprayed their lawns with liquid gold to make the grass grow, just so they could burn black gold to cut it down on the weekends. Our children and grandchildren will wonder why we were so dumb."[9]

So, what's the one thing that will define how new housing developments look and how existing cities and suburbs may get retrofitted over the next fifty years? Once again, the one-solution-fits-all era is over. The answer is that some homes will have gardens surrounded by stone walkways and decorative rock or bark; some will keep lawns but irrigate them with the water from their bathtubs and showers; some will install cushiony Astroturf materials for children to play on year-round; others will plant ornamental bushes, cactus plants, or trees that thrive in local climates. Water-short cities like Las Vegas are, in fact, already actually paying their inhabitants to take the grass out of their yards and replace it with rock, cactus, or other native plants that don't require artificial irrigation. In short, suburbs that are currently defined by a monochrome of green will become more diverse patchworks of native plants, rocks, and other materials.

The American love affair with grass will likely always be with us, but we may need to find other ways of admiring it. Bart Giamatti, commissioner of major league baseball from 1986 until his death in 1989, was earlier in his life a classics professor, and he liked to say that the word *paradise* derived from an Old Persian word for an enclosed green space or park.[10] For a time, baseball parks strayed from that ideal; owners installed artificial turf in many ballparks during the 1960s and 1970s. The parks built since the 1980s, however, have been returning to the beauty of grass. Fans seem to appreciate being close to such an amazing bit of carefully tended horticulture; many of us remember the exhilarating experience of emerging from the dark subterranean tunnels of a baseball stadium and out into the sunshine and brilliant green of the playing field. As we shall see with a number of the trends in this book, the trend in baseball is to return to its roots—to stick with tradition, and part of that tradition is grass. It worked for generations to play on grass, and with luck it will work for generations in the future. Thousands of people on any given night will pay good money to watch grown men playing a game on what many think of as a small bit of paradise.

After leaving a game, however, they may not go home to a lawn but instead a landscape with a garden, some trees, and perhaps a few varieties of ornamental grasses. These are not the perpetually neutered 1½-inch blades of green, but plants that explode upward, reaching as much as 12 feet in height with seedpods and leaves of a broad variety of shapes and colors. These grasses not only get to live a full life cycle that their perpetually prepubescent cousins in the turf can never achieve, but they do it while drinking only a tiny fraction of the water.

Right now, the idea of seeing grasses reach their full height in areas surrounding homes may seem weird, but to plant biologists and water providers, it will be the end of what has been a particularly weird chapter in plant history.

❧

[1] Victor Margolin and Richard Buchanan, *The Idea of Design*. MIT Press, 1995.

[2] Frank J. Scott, *The art of beautifying suburban home grounds of small extent: The advantages of suburban homes over city or country homes; the comfort and economy of ... trees and shrubs grown in the United States*. D. Appleton & Co., 1870. Quoted in Michael Pollan, *Second Nature: A Gardener's Education*. Grove Press, 2003.

[3] Bradley S. Klein and *Golfweek* magazine, *A Walk in the Park: Golfweek's Guide to America's Best Classic and Modern Golf Courses*. Sports Publishing LLC, 2004.

[4] Rebecca Lindsey, "Looking for Lawns." Nov. 8, 2005. NASA *Earth Observatory*. http://earthobservatory.nasa.gov/Features/Lawn/.

[5] US Department of Agriculture, "Selected Crops Irrigated and Harvested by Primary Method of Water Distribution, United States: 2008." http://www.agcensus.usda.gov/Publications/2007/Online_Highlights/Farm_and_Ranch_Irrigation_Survey/fris08_1_30.pdf.

[6] National Club Association, "National Golf Day 2010." http://www.nationalclub.org/club/scripts/library/view_document.asp?CLNK=1&GRP=&NS=PUBLIC&DID=98547&APP=80 .

[7] Scott Yates, "Florida Utility Tests Stormwater Retrieval." *AWWA Streamlines* 1:26, December 22, 2009. http://www.awwa.org/Publications/StreamlinesArticle.cfm?ItemNumber=52619.

8 Future Farm Industries CRC, "Salt Tolerant Cereal (FP12)". *Future Farm Online*. 2009. http://www.futurefarmonline.com.au/research/future-cropping-systems/salt-tolerant-cereal.htm.

9 Michael E. Webber, "Energy versus Water: Solving Both Crises Together." *Scientific American*. October 22, 2008. http://www.scientificamerican.com/article.cfm?id=the-future-of-fuel.

10 George F. Will, *Bunts*. Simon and Schuster, 1999.

CHAPTER 3

THE FUTURE OF WATER
USE INSIDE THE HOME

The water consumed inside a home is some of the most scrutinized water usage anywhere precisely because it is in the home, right where we can see it and keep track of it. If the leaky lawn sprinkler head were a dripping sink faucet located where people saw it every day, instead of underground in the corner of the yard, it would likely get fixed a lot more quickly.

Figure 3-1 on page 30 illustrates typical home water usage, both inside and outdoors. So, how *will* we use water in the home in the future? Very carefully.

Although it is not widely recognized or publicized, as a society we already use less water when compared to previous generations. It wasn't that long ago that water use in the home was much more wasteful than it is now. The Pacific Institute—the leading think tank in the water resource arena—has studied this issue closely and reports that per-person usage of water in this country is already on the way down and has been for the past several years.[1] How has that happened? Mostly with efficiencies in large-scale agriculture, but also in the home.

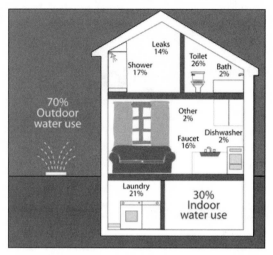

Source: CFPUA.

FIGURE 3-1   WATER USAGE AROUND THE HOME

THE TALE OF THE TOILET

Let's first consider one of the biggest users of water in the home: the toilet. Over the past twenty years, a huge number of the 4- to 6-gallon-per-flush toilets have been replaced with ones that use 1.6 gallons (6 liters). That's progress—but it's still a lot of water when you consider that this is highly treated *drinking* water. This is meticulously treated water, carefully tested, and delivered at great cost to the home in a condition such that a human can drink it every day and live a healthy life. The painful irony of our system today is this: that very same clean drinking water is poured into a bowl, which we proceed to urinate and defecate into, then flush it all away into our sewers.

In his excellent book *Unquenchable*, Robert Glennon relates the story of an academic who welcomed into her home a member of the Tarahumara Tribe of southern Mexico, where water is very scarce. The man had never used a bathroom in the United States, and when he went in he did not know what to do. "In the end, he emptied his bowels in her bathtub rather than her toilet," Glennon writes. "In his culture, water is sacred and no sane person would ever contaminate potable water with human waste."[2]

So will we continue to use our vital water as a central part of our waste disposal system in the future? Possibly. Human waste is foul stuff. It is packed with all the indigestible fibers and food waste, bacterial matter, sloughed-off dead cells, and intestinal secretions that our bodies reject because they are not useful—it's just nasty stuff, and it's dangerous to our health. Just a little bit of *E. coli* bacterial material mixed with drinking water for a short time period can kill even healthy people. Fecal matter is well known to be highly pathogenic, meaning that it causes disease in whatever is around it. So, are there alternatives to using clean water to receive and carry away our wastes? What will toilets look like in the future?

Low-flow toilets have received a lot of attention over the past several years, but in the future we'll see many new approaches. There are currently some incinerating and composting toilets that manufacturers insist work well in specific applications like remote mountain cabins. Right now, such models are uneconomical, impractical, and rare in homes in the cities. However, dramatic improvements lie on the horizon, one way or another. For eons, humans have composted their solid waste, and around the world many people still use latrines dug right into the ground. The problem with this approach, of course, is that we have more and more people often living in concentrated areas. As the planet becomes more crowded, more people are getting sick from waterborne illnesses, often because of improperly located or disposed-of waste. As populations continue to migrate from countryside to cities, the proper disposal of solid human waste will continue to be a more and more pressing problem.

Obviously, this is already a critical public health crisis in many parts of the world—particularly in the megacities where essentially no wastewater infrastructure exists, and where there just isn't enough space and land to use the older, time-honored approaches to human waste management. Take a walk around downtown Accra, Mumbai, or São Paulo if you want to get a sense of just how bad and smelly things can get. The scope of human waste management and disposal issues in many of these megacities is truly daunting. The capability of our oceans to absorb much of this waste is remarkable, but we are rapidly approaching serious environmental catastrophes.

On the other hand, it's possible that toilets may move to using a little bit *more* water, because they will include the function of washing as well as flushing away the waste. Some toilets in the future may work more like a bidet combined with the toilet. This may sound like it's using more water, not less, until you really look inside the water numbers. If you use a toilet, you use toilet paper—and that may be more of a water problem than you realize. Let's step back and look at the overall situation from a broader perspective. It's time to consider in more detail the idea of virtual water and the related term water footprint, which we defined in chapter 1, two important and emerging concepts that we will return to throughout this book.

THE TP FACTOR

A good example of water footprint is the case of toilet paper and how much water goes into its production. Environmentalists bemoan the fact that most of the toilet paper sold around the world contains at least some fibers from old-growth trees that may be three hundred or more years old—and that's just the start of the problems with toilet paper. The process of manufacturing toilet paper uses extremely large quantities of water—and energy—at every stage of the process. The water used in manufacturing may not be treated drinking water, but it's still a lot of freshwater being used up. In addition, the facilities that make toilet paper produce huge amounts of wastewater coming out the back end of their plants. Manufacturers aren't keen on releasing the details, so it's impossible to know for sure, but a group of scientists in the Netherlands that studies the water footprint of various products estimates that a roll of toilet paper made from virgin wood fiber may use somewhere in the range of 75 to 150 gallons of water per roll.[3]

Exact numbers are hard to come by, but several market estimates show that Americans use fifty to one hundred rolls of toilet paper apiece per year. Recycled toilet paper is popular in many parts of the world outside of the United States, but in the United States it only accounts for a small percentage of rolls sold.

If Americans won't switch to recycled TP, maybe they'll switch to using a direct spray of water. The water use from a bidet also varies, but because it's a small stream, most manufacturers estimate that for a family of four it will use a half gallon (less than 2 liters) or less per day. Because no hands are used, the spraying method has also proved to be more hygienic. Toilets with a bidet function are well known and popular in Europe and Asia but relatively rare in the United States. That's something that also may change.

Doing the math, if a family in the United States used a bidet primarily and stopped using toilet paper, the amount of total freshwater consumed by the whole process would be reduced by something like 75 to 90 percent. It's no wonder that environmental activists have pointed out that it's better for the planet to drive a gas-guzzling Hummer than to use nonrecycled toilet paper every day. While in 2010 the Hummer was teetering on the edge of extinction, in part because of environmental concerns, no sharp movement toward recycled TP or bidet use occurred. However, with growing water scarcity and sharply increasing water prices looming over the next few decades, we can also expect all of this to change.

RECYCLING TRENDS

Another big change will likely come in the way that US homes are fitted with plumbing. Right now it's hard to imagine that we might have some kind of separate systems to handle solid human waste, urine, and other water used in the home, but think back just a few years to how we formerly handled trash. It wasn't that long ago that everything went into one trash can, and all of it was carted away... somewhere. Now most homes in the United States have at least one recycling bin, and many have several. In the future, there will likely be a similar division of water and wastewater streams both coming into and going out of the home.

It's instructive to look at what urine is made of. Mostly, of course, urine is water—95 percent or more, depending on how much water a person has been drinking. That's why NASA is now recycling all of the urine and sweat produced by astronauts on the International Space Station. The cost of lifting anything into space today is between $5,000 and $10,000 per pound. Think about that glass of Tang: the powder might get to space pretty cheaply, but a gallon of water costs more than a new car. Given that, it's surprising that NASA unveiled only in 2009 a wastewater recycling system. That the space agency did finally get one working after years of effort made the $250 million investment seem smart, especially because that capability will be crucial for any manned flights that go beyond Earth's orbit.

NASA has been the driving force behind the development of lots of things now in use in the home. And the NASA system for urine recycling would actually be much easier to use on earth than it is in space. At the heart of the system on the International Space Station is a simple distiller that boils liquid and captures the steam. On earth, gravity can pull the stream droplets down into a collector; in space they need a centrifuge that can spin steam around and collect the droplets. Even if the price for a urine recycler dropped to one ten-thousandth of its current price of $250 million, $25,000 is still a hefty price tag for a family to spend on a device that does something that makes most people wince when they hear about it. But don't discount the concept.

Even if we don't turn urine into drinking water, we could turn it into fertilizer. Of the 5 percent of urine that isn't water, most of what's left is water-soluble and contains nitrogen, potassium, magnesium, calcium, and phosphates. If that list sounds familiar, that's because it's exactly the same as the stuff that people buy every day to fertilize the lawns and gardens that we just discussed. The popular NPK (nitrogen, phosphorus, potassium) rating for fertilizer is the balance that many gardeners tinker with endlessly, trying to grow greener grass, prettier roses, and bigger melons. They may just need to look to their own bathrooms for the proper recipe. Researchers at the University of Kuopio in Finland collected urine during the winter of 2007–2008 and stored it in a refrigerator until growing season. They then grew several rows of crops and found that the urine mixed with wood ash produced more tomatoes than plants treated with commercial fertilizers. And a double-blind panel of taste-testers said the urine-fertilized plants produced better-tasting tomatoes.[4]

Again, think about our trash streams. Once it was a single stream, and gradually it's turned in to a multitude of streams. While retrofitting a home with three sets of waste pipes may seem extreme now, it may seem routine in the coming years. Indeed, at some point in the future, our descendants will almost certainly look back on this era—carrying our debris away from our homes with clean water—as one of profligate waste.

But before all those changes can come to pass, there is one issue that will need to be faced one way or another: the smell. Especially the smell in public.

One current innovation involves waterless urinals. These devices have various contraptions for draining urine away from the bottom of the bowl and trapping the accompanying odors. The problem is that they don't always quite do the job of eliminating the odor. Brand-new buildings can end up smelling like a New York City subway station as a result of waterless urinals. Manufacturers disagree with this assertion and claim to have made great strides in controlling the smells. They have come a long way, but they still have a long way to go. The companies that can dispose of human waste and actually figure out how to neutralize the odor are the ones to watch.

They face a hard challenge. Water, in addition to being the essence of life, is just exceptionally good at carrying away and sealing off our waste. People often have no idea how smelly their own excrement is until they go camping or have to produce a sample in the doctor's office. Human waste mixed with water and then flushed behind the water-filled barrier of a gooseneck trap in a toilet is rendered instantly and effortlessly invisible to the nose and eye. Water now does the job so well that our standards for the elimination of smell are set very high. But have no doubt about it—major changes are coming to our plumbing.

The constant flushing of toilets is the biggest user of water in the typical household. As we've seen, this is likely to change. After that comes water used for washing our clothes, washing ourselves, washing our dishes, and finally—last, least (in terms of volume) but most important—water used for cooking and drinking. Let's examine all these uses and also look at graywater systems.

CLOTHES WASHING

Manufacturers of clothes-washing machines, like manufacturers of toilets and showerheads, have made great strides in terms of lowering water (and energy) consumption. The US Environmental Protection Agency (USEPA) estimates that, for the most popular models in use today, the average machine uses between 3,000 and 10,000 gallons per year in the United States[5]—a lot of water, but down from the consumption rates of older machines.

Anyone who went to sleep in the year 1960 and woke up today would find many things about life to be radically different, but one thing that hasn't changed much is the washing machine. Front- and top-loaders were around then, and the way the clothes got washed hasn't fundamentally changed much. There is little chance that this will still be the case in 2050. Washing machines will look and act very differently.

For instance, an inventor from Turkey is promoting his Washup design,[6] which incorporates a washing machine and a toilet. The space above the water storage for the toilet is occupied by the front-loading high-efficiency washer. The unit then stores the water that was used to wash clothes and later uses it to flush the toilet. A little different from what we are used to, but it makes sense. Who cares if the water used to carry away our waste has a little soap and dirt in it?

While manufacturers have made some big advances—reducing by about half the amount of water consumed for clothes washing—much larger steps will be taken in the future. One candidate is a washing machine from a British company, Xeros, that uses small plastic balls and just a cup of water per load.[7] A bit of humidity and the right kind of static charge, along with some tumbling, are enough to properly clean a load of clothes, says the manufacturer. The unit is about the size of a regular washing machine.

A couple of other manufacturers have developed similar prototypes using some form of resin as a method of removing smells, stains, and dirt from clothing. If one of these can be proven to work and can be manufactured on a scale large enough to serve people and businesses around the world, it will translate into huge potential water savings. Much of the water used in even the super-efficient front-loaders is actually used to remove the soap from the now-clean clothes. That is, the soap gets into the clothing to clean it but then stays behind, and the only way to remove the soap is to use gallons and gallons of water to rinse it out. That's why some of the washers of the future are being designed to clean without using any detergent at all. Another idea that seems to make sense—if it works.

The Airwash, invented by two students from Singapore, won an international competition for its conceptual design, although it's not available for sale yet.[8] The unit is taller than a regular clothes washer and about the same width but only a few inches deep. It uses negatively charged ions and compressed air to clean clothes. The concept is that it would clean one article of clothing at a time so that a person would theoretically clean clothes every day, avoiding the need for keeping dirty clothes in a hamper until wash day. Even the inventors say this design is at least ten years away from being mass produced, but if it catches on, it could work in any room of the house and eliminate the need for plumbing to wash clothes, not to mention all the water used.

Other new clothes-washing devices abound, including one from Electrolux that uses ultraviolet light and free radical oxygen ions to clean clothing.[9] No water at all in this one, either, but it has one significant problem: It will only clean clothing that has a particular nano-coating with a material similar to Teflon, unfortunately something that is not available for sale right now. The Electrolux Group is betting all clothing in the future will have some kind of nanocoat, so it is trying to get out front with a machine to clean such clothes.

While these kinds of ideas may at first seem a bit far-fetched, they're really not. This last idea would not only save the water needed to wash clothes—it would also save the much larger volumes of water needed to grow the cotton to make those clothes in the first place. (We'll explore that concept more in the next chapter.) Nano-coatings that make other artificial fabrics feel like cotton may be common in the future, so a waterless cleaning machine may make all kinds of sense.

One other decidedly low-tech approach is that people will just wash their clothing less as water becomes more and more expensive. We could move in this direction quite a ways before body odor became an issue. Americans wash their clothes far more frequently than people in other countries, something made possible at least in part because of how cheap water is here. As with so many of these activities, the question to ask yourself is this: if the price rose, would your behavior change? Let's say you pay, on average, $40 a month for water inside the home, and you know that about a fourth of that, or $10, went for washing clothes. What if the price doubled? What if it were ten times the price? If you paid $200 or $400 a month for the water needed to wash your clothes, would you maybe wear a pair of jeans a couple of extra times before you washed them? Think about it, because water prices are definitely going to rise.

The economics of making some of these investments also changes with the price of water. A washing machine that uses a fourth the amount of water but has a sticker price that's twice as high makes almost no sense economically today for most families. Appeals to save the environment work sometimes with some people, but in general people make purchasing decisions based on their wallets, especially during a tight economy. Right now, the cost of water is almost absent in such a calculation—but it won't be in the years to come.

BATHING

The bath is a luxury that humans have enjoyed for a long time. Warm baths, all the better. Historically, wherever hot springs were discovered, communal baths were quickly established. Such is the enduring popularity of the hot bath. It's pretty hard to find a home in the United States without a bathtub. A home bathtub, however, is still not common in many other parts of the world. Nations with more money and access to water have more of them. Americans, as is so often the case, lead the way in terms of guzzling water for baths.

Many US water utilities have done yeoman's work in encouraging people to take showers instead of baths, to install flow-reducers into showerheads, and to take shorter showers in order to reduce overall consumption. Recent per capita water consumption decreases have been dramatic in Western cities like Denver and Albuquerque. These efforts do make a difference, but they rely on people being educated and willing to reduce water use on a voluntary basis. A more thoroughgoing cultural change will eventually be called for. Recent droughts have accelerated this type of cultural change already in Australia, where short showers every day are encouraged. There, for example, radio stations play four-minute songs in the morning to help people better time their showers.

Technology may also help ride to the rescue here. Antiperspirants and deodorants have improved some over the past fifty years, but antiperspirants still use some variant of soaking aluminum salt into the skin, while deodorants just smell better than your average underarm. Scientists are working on new technologies that avoid the downsides of coating part of the body in aluminum but that essentially trick the nose of anyone nearby into thinking that no bad smell is emanating from a person's armpits.[10] Modified foods of the future may also contain nanoproducts that can reduce odors from perspiration. In other words, maybe we won't need as many showers or baths in the future.

The whole history of bathing and cleaning, like many of the other stories in this book, took a dramatic turn after World War II. Returning soldiers who were used to spending many weeks without bathing became fans of washing daily, especially as they moved from farms with water often heated on a stove to suburban homes with hot and cold running water. And, of course, the emergence of nonstop advertising for soaps and shampoos helped shape several generations into thinking that daily bathing was essential for proper hygiene. Now that those habits have become enshrined in the DNA of American life, it may be difficult to turn back the clock. But it is possible.

A small but growing movement of people is pointing out the health and environmental detriments of soap, especially shampoo. You may not have heard of the No-Poo movement—but adherents actually do call it a movement. They point out that the evidence that shampoo is bad for your hair is sitting right next to the shampoo: the conditioner. Those who don't wash out all of the natural oils from their hair don't need to replace the moisture with artificial products.

Unwashed hair doesn't become greasy, but it is noticeably less frizzy, advocates say. A similar movement encourages people to reduce the number of times they bathe, shower, or wash up, in part because we can damage our skin with overly frequent washing and the use of an endless variety of artificial chemicals.

And it's not only an issue of the effect of all these chemicals on our skin and scalp. These various health and beauty products are also a prime source of the very low-level contaminants—increasingly referred to as xenobiotics—that are now routinely flowing through our water systems, disturbing or damaging our endocrine and reproductive systems, and perhaps irreversibly contaminating much of the planet's water.[11]

At the same time as we face pressure to reduce water use, however, we face a rise in concerns about cold, flu, and other bugs that can be transmitted person to person, especially so-called super bugs that are resistant to treatment with antibiotics. This is a genuine health risk, and it's why our mothers told us to always wash our hands. For this combination of reasons, hand sanitizer has recently become ubiquitous in public places. But this solution also carries with it a raft of concerns, from the deceptive nature of some of the advertising to the drying effects of constantly rubbing a solution onto our skin that is made up of two-thirds alcohol.

Health experts continue to say that the best way of avoiding germs is to wash with soap and lots of warm water. Unlike with washing machines, no significant technological advances appear on the horizon that might change how we wash our hands in the future. Water is exceptionally good at cleaning human bodies, and we should probably count on that continuing to be the main medium for human cleaning in the future. Water is, after all, often referred to as the universal solvent. Because we will probably be washing ourselves with water for many centuries to come, many inventors and water-reuse advocates have been creating systems for capturing and somehow reusing that water—often referred to as *gray water*—to water gardens or flush toilets.

GRAY WATER SYSTEMS

In the United States, several manufacturers are marketing gray water systems for homes—systems that capture water from showers and baths and use that water to flush toilets or for other purposes. In some cities, especially in California, such systems are legal, but they remain a violation of local building codes nearly everywhere else. Advocates for these types of reuse systems sometimes encourage people to violate the codes even where the systems are legal because the bureaucratic requirements are so burdensome. Stories of slow-moving bureaucracies are widespread, especially of municipal departments that approve changes to plumbing codes.

No sweeping changes are being planned by the International Association of Plumbing and Mechanical Officials, the group that establishes the uniform code on which most municipalities base their own codes. But even that staid organization has approved one gray water device, a small unit manufactured by WaterSaver Technologies of Louisville, Ky.[12] Their system comprises a tank that goes under a bathroom sink and some plumbing, a purification system, and related bits of hardware that allow the tank to fill the toilet reservoir for flushing. It's not a radical idea or space-age technology by any means, but it's one of a large number of choices that will save water in homes in the future.

As we will see in later chapters, all areas of the country have water conservation and consumption issues that are creeping up and that will have to be confronted in the coming decades. However, some states are dealing with such issues more aggressively now. Nevada has paid people to take out their grass in order to save on watering charges.[13] And in Tucson, Ariz., a city ordinance passed in 2008 requires that, as of the summer of 2010, all new commercial buildings must install a rainwater harvesting system and have a plan to show how the system will be able to meet 50 percent of the irrigation needs for the landscaping.[14] (Golf courses are exempted from this new ordinance—naturally.) It's a radical step forward for a city of more than a half million people, but it will likely be a model for dozens or even hundreds of other cities over the next fifty years.

If the plumbing and building codes don't begin to change with actions like those of the city council in Tucson, we can expect that court battles may force the hand of local governments. It's not hard to imagine this scenario: a local blog writer does a story complete with pictures about a gray water reuse system he installed in his home. The blog item gets picked up by a local TV station, which broadcasts a story with video of the pipes feeding from a bathroom to a backyard garden. The reporter from the TV station then asks the city if they have any comment, one thing leads to another, and the planning department issues some kind of order against the blogger, insisting that he remove his gray water system. He refuses and gets representation from any number of environmental law firms interested in the publicity and in setting a precedent. The blogger loses at trial, but he appeals and after a while a higher court will find itself having to decide between the rights of a municipality to protect the environment as it sees fit versus the rights of an individual to protect the environment as he or she sees fit.

Water law has been a growth industry for many years, but it has largely been confined to interstate compacts, deals between farmers and increasingly thirsty cities in the West, and disputes between different states or between states and the federal government. In the future, water law disputes may define how water is used down to details as seemingly small as the way in which a person can collect or use water from his or her own toilet or bathtub.

DOING THE DISHES

In terms of water use in the home, dishwashing ranks below bathing, but still uses a substantial amount of water. The key point here is that we will continue to need to wash our dishes. Just as we discussed with toilet paper, the true water use, or water footprint, of manufacturing paper plates or other disposable dishes is quite high. There are also negative impacts on the environment, as all this stuff has to be thrown away after it's used. Furthermore, even the truly biodegradable components of disposable dishes have a significant environmental cost because they are made most often from corn, which itself requires a lot of water, not to mention petrochemical-based fertilizers, to grow. People have been eating from glass and clay plates cleaned with water for thousands of years now, and we don't expect that to change anytime soon.

The cleaning of those plates does not take huge amounts of water, comparatively speaking. Nonetheless, new dishwashers are getting better and better all the time at cleaning dishes thoroughly using high-powered pumps and special sensors that can detect if dishes are actually clean by optically scanning the water that flows through the pumps. The real trick to saving water here is consumer education: the new washers are so good that manufacturers insist that it's a waste of water to rinse off dishes before putting them in a dishwasher. Let the dishwashers do their job, the makers say. Dishwashers will certainly use less water in the future, but likewise, they will probably cost more. Just like with clothes washers, the rising price of water may eventually be the thing that convinces many families to invest in a new dishwasher that uses less water.

Times change and people change, and over time, peoples' behavior will change, too. Our parents' generation used to wash all the dishes pretty thoroughly *before* they put them in the dishwasher—partly because they didn't believe the "darned machine" actually worked (which it may not have) and partly because the water used was basically free. Our grandchildren won't be doing that.

Several patents have also been issued for various types of ultrasonic dishwashers. Most jewelers use an ultrasonic bath right now to make jewelry shine. Dishes are submerged in water in the sink and shaken with a very small but very fast shaking motion to loosen dirt. Further down the road, there may be dishwashers that can clean and sanitize dishes using only bursts of air infused with steam. One such unit, the Rockpool, would then sanitize dishes with ultraviolet rays in a trim countertop unit.[15] The whole process uses very little water but probably would use more electricity. Electricity has its own (extremely high) water demands, so this is no water-saving panacea, either, but if water becomes much scarcer, a dishwasher like this could be in high demand. This particular kind of design is also being developed by Electrolux—the same company behind the clothes washer that only washes clothes coated with nanoparticles. This dishwasher will also require a cultural shift, as it only washes one or two dishes at a time, but the ability to wash dishes with nearly no water may make such a change more attractive in the future.

COOKING AND DRINKING

So that brings us to the last section of this chapter—the most critical, but also the smallest use of water in the home: drinking and cooking. That's right, the most critical use of all, drinking, typically represents less than 1 percent of our total domestic consumption. But what an important percent it is. In the United States, that water is delivered to us safely every day for a fraction of a penny per gallon. As the water utilities and many public health advocates are wont to point out, this really has to be considered one of the great economic bargains of all time.

Once we add in that 1 percent, we've pretty much summed up the situation in terms of our water usage around the home. But the story doesn't quite end here. A couple of huge and far-reaching conundrums involving drinking water haven't yet been touched on.

First, although our drinking water is treated to very exacting standards and is generally very safe, many people argue that it may not be safe enough. All public drinking water systems are regulated by the US Environmental Protection Agency, which requires regular testing for almost a hundred different contaminants. Even so, many environmental advocates suggest that we should be testing for and removing dozens of additional potential contaminants. They say that for the United States to continue to provide safe water in the future, the utilities will have to figure out how to handle a plethora of additional issues such as pharmaceutical residues, by-products of medical drugs, personal hygiene compounds, and a broad array of other contaminants that aren't treatable with chlorine, membranes, ozone, or the other methods traditionally used by water providers. Given all of these new and often only recently discovered contaminants in our drinking water, it's going to be tougher for utilities to continue to provide safe drinking water under more and more demanding regulatory standards in the future.

The second conundrum results from the fact just mentioned: *very little of our treated municipal water is actually used for drinking.* All of this water usage that we have been discussing—from the flood drenching a vast green carpet outside the home, to the flushing and washing done inside the home, to that tiny bit that we actually consume—is accomplished with highly treated water. In a municipal system, *all* of this water is treated to high US government standards and is safe for any man, woman, or child to drink. So the obvious question is this: if we only drink 1 percent of it, what sense does it make to treat all the rest of that water to the same exacting standards? Can we really afford to treat all the rest of that water? We'll write more about what this whole issue might look like from the perspective of the water utilities in chapter 8, but from the perspective of the home, the answer isn't at all clear.

Many factions in the water business argue that we shouldn't treat our water in huge and expensive centralized facilities but instead on a more local or *distributed* basis. In other words, we should treat water at the point where we are going to use it, and we should treat it to the appropriate level of quality for the specific and intended use. If we are going to drink it, then we should treat it pretty thoroughly; on the other hand, if we are going to water our lawn with it, then maybe it doesn't need to be quite as clean.

There is a whole sector of the drinking water industry called the point of entry/point of use (POE/POU) treatment sector. This sector manufactures and sells a broad array of products geared to just this concern. This includes the various home treatment systems that more and more people are installing in their basements to further treat the (already treated) municipal water that flows into the house, the under-sink carbon filters and UV systems used to treat kitchen water before it is used to drink or cook with, the carbon canisters on the backs of our refrigerators, and right down to the pour-through filter devices on the water pitchers inside the refrigerator.

This question of centralized vs. decentralized treatment is a huge and controversial issue in the water business, but it is clear that a truly distributed system of treatment would require us to completely redesign and rebuild our entire water treatment and distribution infrastructure. Hence, widespread distributed treatment is not likely to happen any time soon. But the trends are rapidly moving in that direction, with, for example, separate potable (drinkable) and recycled wastewater collection lines being required in new homes in certain parts of the country. From a long-term perspective, and since drinking water regulatory standards will get tougher, it seems clear that we won't be able to treat our municipal water to these more exacting standards forever. An inexorable move to more logical and less expensive decentralized systems will be one of the key changes that our children and grandchildren are likely to experience.

BOTTLED WATER

Another key issue in the discussion of how best to provide clean drinking water to the home involves a trend that nobody in 1960 could have seen coming: the explosive growth of bottled water (see Figure 3-2). Now, fifty years later, Americans purchase more water— nearly all of it in plastic bottles—than beer, juice, or coffee. The only liquid with higher packaged sales as of this writing is soda, but even that may not last much longer, as bottled water has been chipping away at soda consumption numbers for the past ten years or so. Will this trend continue? Here, there is little consensus. In fact, the bottled water question is perhaps one of the most contentious issues found anywhere in the entire water business, and so we will take a little detour to review this issue before we wrap up the chapter.

At the same time that many parts of the world face crippling water shortages, it seems outrageous that the bottled water craze continues to captivate wealthier regions. Hollywood starlets still pitch all manner of natural spring waters, vitamin waters, energy waters, smart waters, and various other so-called specialty beverages right up to Bling H_2O, which proudly calls itself the most expensive bottled water available.[16] These bottles of water are all now available at a cost of only a few hundred to a few thousand times the price of the tap water from which they are virtually indistinguishable. Liquid OM is, according to *Newsweek*, a "super-purified bottled water containing vibrations that promote a positive outlook." The water purportedly

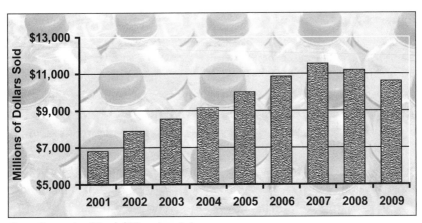

Source: International Bottled Water Association.
FIGURE 3-2 BOTTLED WATER SALES IN THE UNITED STATES

possesses an energy field made by striking a giant gong and Tibetan bowls in the vicinity of the water. "The good energy can be felt not just after you drink the water but also when you're just holding the bottle," the bottler told the magazine.[17] It makes you wonder what can possibly be next—and it calls to mind the famous quote from H.L. Mencken that "no one ever went broke underestimating the intelligence of the American public."

Sure, bottled water is generally safe, because the giant corporations selling it understand that they are promoting the perception that their water is safer and tastes better. They know full well that even one story of someone getting sick or dying after drinking bottled water would pop the balloon of that perception, and sales could suffer. For example, remember the crisis that Perrier faced a few years ago when traces of benzene were found in the iconic French bottled water? As for taste, well, there are no regional variances in, for instance, Dasani, which is produced by The Coca-Cola Company. The local bottling plants use local water, often from municipal sources, and then filter, treat, or boil it the way they do to make sodas. Coke then adds in some minerals at precise levels, so it tastes the same if you buy a bottle in Boston or Albuquerque. PepsiCo sells Aquafina with no mineral additives; it just treats the water from whatever the source and sells it with nothing added. And at least this water doesn't have to be shipped over an ocean, as in the case of Evian or Fiji Water, using up large amounts of energy in the process.[18]

And let's face it: bottled water is probably better for you than soda pop. Some health experts have suggested that while tap water cannot compete with sodas or other high-sugar drinks for children's attention, perhaps bottled water can. And, in light of the urgent need to combat childhood obesity in this country, bottled water should be made available to kids. Right now, food stamps can be used to purchase bottled water, and those purchases are encouraged if it cuts down on less healthy alternatives.

The environmental concerns around bottled water are legion, as water resource expert Peter Gleick of the Pacific Institute has pointed out in his recent book *Bottled and Sold: The Story Behind Our Obsession with Bottled Water*.[19] The vast volume of plastic bottles—many of which are now floating around in the North Pacific Ocean—and the fossil fuels used to create and transport this water around the world (when gravity serves most public systems just fine) are only two of the problems. Additionally, the bottle itself raises issues: the science here is still young, but some studies have shown carcinogenic attributes in nearly all the different kinds of plastic used. A few industrialists, typically from Iowa or Nebraska, have been promoting the use of bottles made from corn by-products, but the water footprint and the energy used to grow that corn and process it into plastic is immense. The list of concerns from environmentalists goes on.

The explosive growth of the bottled water industry over the past couple of decades is a spectacular example of how customer perceptions—rightly or wrongly—and Madison Avenue can create and drive new markets. A lot of Americans have either been convinced that their public water is not safe to drink or that drinking bottled water is somehow cool. We now spend more than $10 billion a year on this stuff. According to the Beverage Marketing Corporation, the reasons are clear: Bottled water is a healthy and safe, ready-to-drink commercial beverage, which is becoming increasingly affordable.[20] Providers continue to market bottled water by promoting it as something completely different from tap water, and that concept clearly sells to a large swath of the American public.

The truth is that little evidence exists to suggest that bottled water is much different from, or safer than, tap water. Also, bottled water is just not as closely regulated as tap water. One of the main reasons for the popularity of bottled water is its transportability. However, this is obviously easily accomplished by keeping a couple of empty bottles around and filling one from the tap on your way out the door.

The bottled water fad may be moderating; some upscale restaurants are now promoting the virtues of tap water. A couple of years ago, PepsiCo laid off some 3,300 people, attributing it to a slowdown in its bottled water business.[21] And no less a water authority than the National Association of Evangelicals has said, "Spending $15 billion a year on bottled water is a testimony to our conspicuous consumption, our culture of indulgence…. Drinking bottled water may not be a sin, but it sure is a choice."[22] With the economic hardships that many American families are now experiencing, it seems likely that fewer people will be willing to shell out two bucks for a drink of the same water that comes out of their taps virtually free.

To be fair, there are also some considerable upsides of bottled water. In many parts of the world—areas with dysfunctional public water treatment and distribution systems, or none at all—bottled water may be the only way for people to ensure that they are drinking something reasonably safe. In these areas, which include much of the developing world, properly treated and packaged bottled water can play a critical role in human health. Unfortunately, it's often the poorest people who have to pay the most (certainly on a percentage of income basis) for this water. Bottled water providers also play a critical role in disaster or emergency relief projects following events like earthquakes or floods, in places where the existing water infrastructure may have been overwhelmed or destroyed. Properly treated bottled water can be a real lifesaver in these sorts of situations.

So, will bottled water play a part of drinking water's future? Undoubtedly. While this system of delivering drinking water is certainly less efficient than delivering municipal water, it's not a broken system. The water that we drink is crucial to everyday life, and it may turn out that it's just not possible to treat all of our municipal water to safe standards, and hence utilities will have to look to other alternatives.

It's not science fiction to imagine a scenario like this: some new contaminant will be discovered (remember, few water utilities worried about *Cryptosporidium* before 1990). Suppose that this new contaminant can't be removed without hugely expensive additional treatment, and yet the USEPA rules that all water intended for drinking be free of this contaminant. A gut-wrenching period will follow in which water utilities essentially give up on trying to provide one stream of potable water for all uses. They keep the flow of water going through their current infrastructure for bathing, washing, and household irrigation. At the same time, the utilities tell the citizens that they should start using more highly treated water or bottled water for drinking, perhaps their own or perhaps water they produce in conjunction with existing bottlers using returnable, reusable bottles. It's not a certainty that a scenario like that will happen, but it would be shortsighted to say it would not.

Given all of this, what's the future of drinking water in the home? Because the amount of water needed for actual drinking and cooking is such a small percentage of the water needed in the home, there's a good chance that in the future, this component will be somehow treated and delivered separately. Perhaps it will be treated at the point of use to more exacting standards or maybe it won't come through the existing municipal system at all, and the trend toward bottled water will only keep growing. It's hard to say, but we shouldn't ignore the role that separately delivered drinking water might play in the future.

❦

[1] See Pacific Institute, "US Per Capita Water Use Falls to 1950s Levels: Analysis of USGS Data Shows that Efficiency is Effective, Demand is Not Endless." http://www.pacinst.org/press_center/usgs/, March 11, 2004, in which the Pacific Institute reviews Susan S. Hutson et al., US Geological Survey Circular 1268: *Estimated Use of Water in the United States in 2000*, http://pubs.usgs.gov/circ/2004/circ1268/.

[2] Robert Jerome Glennon, *Unquenchable: America's Water Crisis and What to Do About It.* Island Press, 2009.

[3] See Water Footprint Network, www.waterfootprint.org. See also for example Arjen Hoekstra, "A Comprehensive Introduction to Water Footprints,"

2011, at http://www.waterfootprint.org/downloads/WaterFootprint-Presentation-General.pdf, or A. Ertug Ercin et al., "Corporate Water Footprint Accounting and Impact Assessment: The Case of the Water Footprint of a Sugar-Containing Carbonated Beverage," *Water Resource Management*, Oct. 23, 2010.

4 Rebecca Boyle, "Better Tomatoes via a Fertilizer of…Human Urine?" *Popular Science*. 2009.

5 US Environmental Protection Agency and US Department of Energy, ENERGY STAR Qualified Clothes Washers. http://www.energystar.gov/ia/products/appliances/clotheswash/Qualified_CW_2010_Criteria.xls.

6 Core 77. Sevin Coskun's WASHUP. Core77's Greener Gadgets Design Competition 2008. http://www.core77.com/competitions/GreenerGadgets/projects/4609/.

7 Xeros Ltd. http://www.xerosltd.com/.

8 Emily Pilloton, Airwash Waterless Washing Machine. Aug. 22, 2007. *Inhabitat: Design Will Save the World*. http://inhabitat.com/airwash-waterless-washing-machine/.

9 Electrolux Group. http://group.electrolux.com/en/. See also Electrolux Design Lab. http://www.electroluxdesignlab.com/.

10 See Karl Laden, *Antiperspirants and Deodorants*. CRC Press, 1999.

11 See for example Glenn Braunstein, "Prescription Tap Water: What Drugs Are We Taking With Our Drugs?" *Huffington Post*, Jan. 19, 2011, http://www.huffingtonpost.com/glenn-d-braunstein-md/prescription-tap-water-wh_b_809870.html; or Despo Fatta-Kassinos et al., *Xenobiotics in the Urban Water Cycle*, 2010. See also Environmental Working Group, www.ewg.org.

12 WaterSaver Technologies. http://www.watersavertech.com/.

13 Dave Downey, "Western Pays Cash to Take Out Grass: 1,500 Murrieta Customers Targeted by Conservation Program. *North County Times–The Californian*. Feb. 23, 2010. http://www.nctimes.com/news/local/swcounty/article_c1f881e3-8682-51be-8920-6d88dba35010.html.

14 City of Tucson Office of Conservation and Sustainable Development, "Business Tools: Green Building and Smart Growth." http://www.tucsonaz.gov/ocsd/business/building/

15 Levent Ozler, "Electrolux Rockpool: Waterless Dishwasher." Dexigner. Nov. 23, 2004. http://www.dexigner.com/news/3239.

16 Bling H20. http://www.blingh2o.com/.

17 Lisa Miller, "Bless This Bottled Water." *Newsweek*, Dec. 8, 2007. http://www.newsweek.com/2007/12/08/bless-this-bottled-water.html.

[18] Elizabeth Royte, Bottlemania: *How Water Went on Sale and Why We Bought It.* Bloomsbury, 2008.

[19] Peter H. Gleick, *Bottled and Sold: The Story Behind Our Obsession with Bottled Water.* Island Press, 2010.

[20] See publications of Beverage Marketing Corporation at http://beverage marketing.com/?service=publications&section=bottledwaterus&hl=bottle.

[21] Andrew Martin, "Tap Water's Popularity Forces Pepsi to Cut Jobs," *New York Times*, Oct. 14, 2008. http://www.nytimes.com/2008/10/15/business/15pepsi.html?_r=1&ref=indraknooyi.

[22] Miller, "Bless This Bottled Water."

CHAPTER 4

THE FUTURE OF AGRICULTURAL WATER USE

Most discussions about world water problems begin with some reference to the fact that 97 percent of the world's water is in the ocean and hence too salty to drink or use on crops. They go on to say that of the 3 percent left, most of that—the numbers start to vary here—is tied up in the polar icecaps and permafrost. The tiny remaining amount is contained in the rivers, lakes, and aquifers that we humans can draw upon to support our needs (see Figure 4-1, page 54). Those writers are all correct. If all the water in the world fit in your house, available freshwater would scarcely fill the bathtub.

ABUNDANCE, SCARCITY, AND COST

Still, that figurative bathtub holds a *lot* of water. The size of the oceans is simply hard to fathom for us on land. Indeed the phrase *hard to fathom* itself has its roots in the nautical world, meaning that it's difficult to gauge how deep the ocean is, or how many *fathoms*— a measure as long as a man's outstretched arms. Now, with modern instruments, we can figure it out, and the oceans are unimaginably big. More than 70 percent of the surface area of the earth is ocean, and more than half of the ocean is at least two miles deep. If you picked up the tallest mountain above sea level, Mount Everest, and

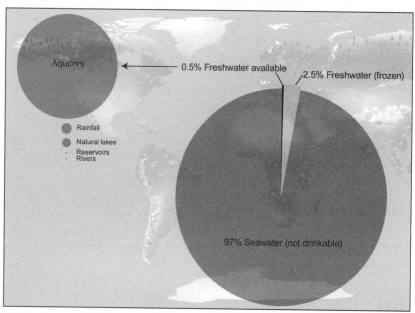

FIGURE 4-1 WATER ON THE EARTH

dropped it in the deepest point below sea level, the Marianna Trench, it would be covered by about a mile of water.

So, the percentages that all the books mention can be a little deceiving. Even though the percentage of freshwater is relatively small, a lot exists on this earth, and there is plenty enough to give every man, woman, and child on earth all they could possibly drink every day many times over.

So, what's the problem? Well, the water isn't always where we need it, and the fact is that humans cannot live on water alone. We all need food, too, and the rub here is that to grow our food, we need water, lots and lots of water. The numbers vary from region to region, but in general, for every single gallon of water that we use in our daily lives, the farmer who feeds us is using lots of gallons. Seeds need dirt and sunshine, but the one thing that really turns seeds into salads is water—and lots of it. In short, agriculture is by far the biggest user of water on the planet, typically using some 70 to 80 percent of all water consumed in most areas.

Right now, the cost of all that water is essentially hidden. Farmers generally pay almost nothing for their water, for one of three reasons: first, because in many regions that water simply falls out of the sky onto their fields; second, because they can pump the water out of an aquifer below their fields without paying for it, and they don't have to replace it; or third, because they get their water from a government-built reservoir or other subsidized water project.

One of the reasons that some foods in the grocery store cost more today is that they involve more processing. The cost of the water that went into producing that food item is scarcely a factor—at least not today. That was the way of the world in 1960, and it is still generally the way of the world today. However, by 2060—and more likely well before that—the cost of water will have much more significance and influence on the price of the food items that we buy.

Consider that of the three types of water input for agriculture (rain, underground aquifers, and irrigation projects), two of them are gradually going away. First, many aquifers are being irreversibly depleted. Groundwater that may have taken eons to collect is being exploited at unsustainable rates by agriculture in many places around the world. Increasingly, we refer to this process as the *mining* of water and refer to this source as *fossil water*—in the sense that it really is a nonrenewable resource, just like fossil fuels. Once we use that groundwater, it is, for all intents and purposes, gone; it may take hundreds, thousands, or even millions of years for that aquifer to recharge. Second, as we will see, many of the major federal dam and reservoir projects that support large-scale irrigation will eventually come to the end of their useful lives, and availability of irrigation waters may decline. (There are some exceptions, of course, but in general that's the case. We'll have more on that in chapter 7.)

So, for those farmers lucky enough to live in the state of Iowa, the Argentinean state of Corrientes, the Indian state of Kerala, or various other naturally wet spots around the world, the cost of water will continue to be essentially zero, even if they have land that's not ideal for farming. But for farmers who aren't fortunate enough to live in high-rainfall areas, the cost of water is going to be increasing. Just as the cost of oil makes a huge difference in the transportation world, the rising cost of water is soon going to change everything about the food business.

How will it change? Well, as water begins to have more than a trivial cost, the amount of water that goes into food will increasingly be reflected in the price. The unavoidable result will be that prices of food will be going up—and the degree to which they'll rise will increasingly depend on how much water it takes to produce the food. The higher the water intensity of the plant, the greater will be the upward pressure on price. It will eventually mean that we'll have to start to be more water efficient in agriculture. And ultimately, for example, it may mean that we'll have to shift our consumption habits toward fruits and vegetables, or meats that need less water to be produced. So, let's examine some foods and see how much water goes into them.

THE WATER COST OF RAISING LIVESTOCK

In our chapter about water use around the home, we started with the biggest uses of water and worked our way down; we'll do that here, too. Agriculture has different characteristics in different continents and countries, but in general, there's no getting around the fact that we use massive quantities of water to produce food. And no food needs more water than beef.

Just as we discussed with lawns, the very notion of raising and eating beef comes from the British, but now it is, of course, increasingly popular all over the world. The stories of the cowboy of the American West and the vaquero in Spain are lyrical and legendary. However, the current state of cattle production is not exactly lyrical; typically it is characterized by massive feedlots and what environmentalists and food quality advocates describe as *factory farms*—or what the USEPA refers to as *concentrated animal feeding operations* or CAFOs. Statistics from the National Cattlemen's Beef Association show that the largest 2 percent of feedlot operators produce six out of seven of the cattle that are eventually turned into our hamburgers and steaks.[1] While the environmental, social, and moral aspects of beef production are outside the purview of this book, little doubt exists that beef production, especially beef that is raised in CAFOs—not by grazing over naturally irrigated land—is almost certainly the most water intensive food that we commonly consume today.

The Dutch group Water Footprint Network estimates that we utilize about 2,000 gallons of water to produce a single pound of beef.[2] Scientists working for the beef industry—as might be expected—peg that number somewhat lower, at something like 400 gallons per pound of beef for cattle raised in the United States.[3] Obviously, the science of water footprinting is still in its infancy, and people still argue about the reasons for these types of wide discrepancies, but either way, it is clear that by eating beef we consume significant water resources. See Figure 4-2 for a quick summary of the virtual water content of common food groups. And the key point here is that most of this water is essentially free right now, because cattle are fed from grain grown in areas irrigated by rainwater, water mined from underground aquifers, or water often heavily subsidized by government-run water projects.

Once again, while water is a critical input in the production of beef, it just isn't that critical of a cost in the production of beef today. Consider the fact that organic or free-range beef is much more expensive at the retail level, even though those cows generally eat grass that grows on open ranges—typically unirrigated pastures.

🍺	One glass of beer	=	20 gallons of water
☕	One cup of coffee	=	37 gallons of water
🥛	One gallon of milk	=	53 gallons of water
🧀	One pound of cheese	=	371 gallons of water
🐓	One pound of chicken	=	469 gallons of water
🐖	One pound of pork	=	756 gallons of water
🐄	One pound of beef	=	1,857 gallons of water

Source: Hoekstra and Chapagain 2008.

FIGURE 4-2 VIRTUAL WATER CONTENT OF COMMON FOODS.

The cost of monitoring, moving, and maintaining those animals is what drives up the price of the meat in the stores—often to twice the price of conventionally raised beef. Sure, the feedlot owners have to feed their cattle a lot of grain, but typically that grain is grown on rain-fed fields or with cheap and subsidized irrigation waters. That essentially free water is what keeps the price of beef so low, as well as the price of most of our other food. More expensive food at the grocery store or at restaurants is costly not because of the water needed to produce it but often because of the transportation, packaging, and marketing. That's going to change in the near future.

The beef industry has had to reinvent itself many times, however, and those on the inside say that it can do so again. One thing cattle producers *could* do, for example, is to use feeds that are produced with less water, especially the by-products of other industries. Indeed, that's already happening now. We'll talk more about ethanol later in this chapter, but the process of creating ethanol leaves a by-product that is essentially just corn without the starch. With ethanol production growing at such a sharp rate, a lot of that by-product is left over, and CAFOs have been the best customers of what's known as distillers' dried grains with solubles—a name originally used for the same by-product in beer and alcoholic beverage production. In other words, if you drive your car filled with E25 fuel (25 percent ethanol, 75 percent gasoline) to the store to buy some hamburger meat, you might be consuming the same processed corn kernels twice.

Another way of improving water efficiency in the beef industry would be if cattle could eat less total grain and instead ate feed that didn't demand so much water and irrigated land. As we know, seawater is plentiful, and manufacturers are already turning some seaweed into a kelp grain additive. Right now, this organic additive is just used as a supplement to help control bacteria in cattle. This use is important, because feedlots typically rely on antibiotics to control bacteria, and the presence of any medicines or growth hormones in meat means that the US Department of Agriculture will not allow it to be labeled as "certified organic." So, natural beef producers have looked to this kelp supplement to replace some antibiotics and to use as a natural mineral supplement. Right now, however, even the kelp producers say the production of seaweed as a direct cattle feed material will probably be prohibitively expensive, and isn't that great as a cattle feed anyway because it doesn't have the protein and other constituents cattle need.[4] If, however, researchers are able to modify barley or alfalfa so that they could be irrigated with seawater, then huge swaths of land currently not agriculturally productive at all right now might become verdant, growing cheaper and more water efficient crops to keep the cattle growing.

Other advances will likely come from within the cattle world. Judith Capper, assistant professor of dairy science at Washington State University, said she doesn't expect that changes will come from the lab in the form of genetically modified cattle but through continued improvements in selective breeding techniques that have been going on for dozens of generations. She said there may also be improvements in hormone implants, ionophores (relatives of antibiotics used to accelerate growth), or beta-agonists (offshoots of an asthma medication, which speed muscle development), all of which help cattle to grow and mature more quickly. These methods are not organic or natural in any sense, but those people who promote these technologies say they are environmentally friendly, because every extra week it takes an animal to reach slaughter is one more week that it needs to eat, drink water, and create waste. "We've made pretty good improvements over the last thirty years, and we should keep doing that over the next twenty to forty years," Capper says.[5]

Of course, this issue has another side. More and more people are concerned with the human health impacts, as well as the ethical implications, of so-called factory farming and the raising of animals in ways that nature did not intend. Many environmentalists and nutritionists point out, for example, that the bacteria control issue may actually be caused by feeding cattle things simply geared to quickly fattening them up—rather than things they were naturally meant to eat. A huge backlash of public opposition has sprung up, particularly in Europe, to genetically modified foods or what many consider to be humankind's inappropriate tinkering with nature. The movement toward organic food—even though the definitions and standards here are often complex and misleading—has definitely taken hold in the United States over the last decade. The buy-local movement has also emerged, as our largely urban-dwelling population attempts to reconnect with its historically rural and agricultural heritage. Certainly, more people subscribe to vegetarianism today than did twenty years ago, some for moral or ethical reasons, others for human health and nutrition reasons. And much broader swaths of the public are just trying to cut back on the amount of red meat they consume. These issues and debates are a bit outside the scope of this book, but, unfortunately, as we will see, water efficiency and conservation objectives sometimes conflict with these local or organic objectives when it comes to food production.

The one sort of silver lining to the large amount of water used in beef production, however, is that even small changes in the beef world can make a big difference in terms of water consumption. That is, if a cattle feed could be developed that even just slightly lowered the consumption of grain from irrigated land, it could translate into huge amounts of water freed up for other uses. Similarly, if consumers reduced beef intake by, say, one hamburger a week, the entire water cycle could reap a massive windfall of newly found water. However, that just hasn't been happening. Beef sales have remained roughly steady in the United States, through all the ups and downs in the economy, even through the recession that started in 2008. The only change has been in the popularity of the cuts of meat, with the fancier and more expensive cuts getting swapped out for more ground beef.

And we see the same trend around the world: as people's standards of living increase, they tend to eat more and more beef and other meats. For example, as China's average per capita income has soared over the past two decades, the country has seen a significant substitution of beef for rice in the typical diet, compounding China's severe water crisis, which we will discuss elsewhere.

So, what is the future of beef? Well, the demand for beef will probably continue to grow as the global economy grows and more and more people enjoy improving standards of living. Animal-rights activists have for generations tried to convince us that eating meat is morally wrong, but moral arguments, on a global scale, have never won the day. The best indicator of beef consumption around the world is income. The more a family makes in annual income, the more likely it is that that family will spend some of the money on beef. Study after study in every country in the world has shown this to be true, even in India where buffalo meat is classified as beef. So as the economies of countries like Russia, India, China, and Brazil continue to grow, we can expect that the demand for beef will continue to go up, at least as long as it doesn't grow prohibitively expensive.

And here we return again to the all-important issue of the cost—and price—of water. If the cost of water continues to rise, as demand from thirsty cities continues to transfer cheap water away from irrigated land, then the cost of beef could start to rise disproportionately. If the price of a Big Mac went from $2 to $5 or even $10, a slow-down in consumer demand would undoubtedly follow. Obviously, a powerful industry like the beef business would try to make sure that doesn't happen, even if it dictated a relocation of production facilities around the country or around the world to ensure that beef continues to be available at the same low prices that we enjoy now. But with the inevitable rise of water prices, this can't last forever.

In the future, will all our beef come from cows? Is there a way to grow beef in a lab? This is one of those predictions about the future that's been around a long, long time. Even Winston Churchill predicted in 1932, "Fifty years hence ... we shall escape the absurdity of growing a whole chicken in order to eat the breast or wing, by growing these parts separately under a suitable medium."[6] It's been more than seventy years, but the science for this prospect of in vitro meat is still in its infancy. There's no question that this would save water and other resources, as the only inputs needed would be those required as it were to grow the meat in carefully controlled environments. Such a scheme would obviously reduce the need for land and water for the animals. The question, however, is whether anyone would willingly eat it. We consume pills all the time that are created in a lab, and we get injections of flu vaccines that are grown in a lab, but so far the consumer reaction to lab-grown food has been negative. If some day, however, a lab-grown hamburger costs $3 and a farm-raised hamburger costs $25, the consumer may be willing to give it a try. With the increasing cost of water, that's the kind of choice consumers may have to make in the future.

If water costs do spike significantly enough to begin to drive up the price we pay at the store for beef, but we still don't want to eat lab-grown meat, what might we do instead? People still need protein, and most consumers still want to eat meat. How does the production of other animals compare to beef?

The numbers vary widely—generally depending on the political views of the organization producing the data—but in general beef needs about 5 to 15 pounds of grain to produce one pound of meat. Hogs do better, with about 4 pounds of feed needed to produce a pound of pork. And every one of those pounds of grain costs a certain amount of water to produce. Hence, eating pork has a lower water footprint than does eating beef. In addition, organic farmers will tell you that hogs are also great to have around an organic farm because when they eat weeds, their chewing and digestive processes will kill the weed seeds that may pass through their bodies. With beef cattle, it's much more likely the weed seeds will pass through unharmed, and the weeds will rise again after a cow has been through a field.

The raising of livestock can also consume water in another manner of speaking if the waste materials are not properly managed and disposed of: that is, clean water can be contaminated. This has been a major problem in a lot of the big CAFOs. On the other hand, if the manure that the animals create is properly handled, it can be an extremely useful fertilizer and in some cases can even eliminate the need for chemical fertilizers. So, while cattle or hogs may eat lots of grain, increasing total water consumption, keep in mind that they can produce not just the end product here—meat—but also one of the essential inputs to crop production—fertilizer. On the traditional American family farm, collecting and spreading the manure on the crop fields was once just a routine farm task, and so it may be again.

With today's massive CAFOs, the management of the waste by-products is not simple and is often not handled in the same type of environmentally sustainable manner. In fact, runoff from confined-animal feedlots is one of the prime sources of surface-water and ground-water contamination in the rural areas of this country. However, the growing trend among all farmers nationwide is to use *all* animal waste products, acknowledging their potential as *resources*.

Many developments today are pointing toward a return to the practices of the past. Many years ago, small grain farmers might have kept one or two hogs around the farm, helping to clean up their pastures and process much of their household waste, as well as providing meat that would last through the winter. Nobody is predicting a massive return to small-scale agriculture, but hogs do provide an excellent way to process household trash, while they're in the process of becoming something for us to eat. Families in many less developed areas and cities around the world will often keep a hog, even if they only have a small yard. This doesn't work too well in the United States, where keeping farm animals around the home is illegal in most jurisdictions. Watch for those restrictions in the United States to fall for hogs in the future, the way they are currently starting to fall for chickens and bees.

Chicken and other poultry, long the darling of health experts, may become more and more popular with consumers if significant spikes in terms of agricultural water costs occur. Chickens convert about two pounds of feed into a pound of meat, so if grain prices go up as a function of water scarcity, the price of chicken in the store will tend to go up only about half as fast as pork and about a fourth as quickly as beef.

In terms of other varieties of meat, the future is not at all clear. Some niche meats may be more widely consumed. For example, cable TV magnate Ted Turner is perhaps the most visible advocate for American bison meat. He's been buying up huge swaths of land for bison production and has opened a chain of restaurants featuring buffalo meat called Ted's Montana Grill. He says that the production of bison is a return to the roots of the United States, featuring an animal that is naturally suited to the climate of the western half of the country.[7] He points out that, in terms of water, the bison is able to survive on the scarce grasses that can grow with the limited rainfall on the Western prairies. This great area extends from Canada to Mexico and across the western half of the country. Bison have huge heads that can sweep the snow off the grasses below, and their wide feet do less damage to the land than cows or even sheep. Their numbers are growing again, thanks mostly to Turner and other enthusiasts who enjoy the flavor or the much lower fat content than beef or even chicken. But even given all this, sales of American bison have remained only a tiny fraction of the sales of beef. As the prices of water and energy escalate, the American bison will probably become more popular, particularly in cities nearer to those natural range-lands where bison thrive. However, worldwide it will likely continue to be a niche meat, much like many of the other niche meats that are only locally popular.

And what about fish? Most species of fish are much more efficient at turning feed into usable protein for people, essentially converting nearly every bit of nourishment into meat. Because many species of fish live on plants grown in seawater, they also don't tax freshwater resources the way land-based animals do. The only problem is that, as big as the oceans are, we are still gradually wiping out many of the most common species of fish. Given widespread overharvesting, dams standing in their way, and other man-made water problems, there just aren't as many fish in the sea as there used to be. According to a *Washington Post* report, worldwide marine fish stocks are falling rapidly.[8] One of the scientists quoted in that report said, "Unless we fundamentally change the way we manage all the ocean species together, as working ecosystems, then this century will be the last century of wild seafood." (And although aquaculture, or fish farm-ing, has been held out as cheap way to address Third World poverty and hunger, in many places it has only spawned a whole new range of challenges, including serious water pollution problems.) We won't try to delve in detail here into the manifold problems of global overfish-ing, but it is one more situation in which an essentially free natural resource leads to an abuse of that resource, just like the previously endless supply of essentially free water for farmers.

If scientists, fishermen, and government agencies can begin to agree on better plans for sustainable fishing, then the prospects here could improve dramatically. Getting even one government to agree to moderate its fish harvest in international waters is difficult, and getting several countries to agree will be geometrically harder, but overall food conditions back on the ground will increasingly demand it. As freshwater supplies grow more and more scarce, and as the world's population grows and becomes more prosperous, more of the freshwater we do have will be needed for drinking water and for domestic and industrial uses in our growing cities. If, and when, that drives the price of land-based proteins such as beef way up, there will be increasing and powerful incentives to raise fish and other seafood on a large scale and in a more sustainable way.

So what's the bottom line on meat? Dwindling water supplies will probably change the meat industry dramatically, but the industry has adapted many times over the last century, and it will continue to change to be able to provide safe meat to consumers at the lowest possible cost. As water becomes scarcer, the price of meat will steadily rise, and most of us will consequently start to eat less of it; it will start to be viewed as more of a high-end food. However, predictions that meat production will essentially vanish in coming years are probably a bit far-fetched.

THE WATER COST OF CROPS

As we said at the outset, for every drop of water we actually drink, we use many more drops to grow the food that we consume. The broad range of fruits, vegetables, and other crops that we consume all have their individual, and often unexpectedly variable, water footprints, and we'll take a look at a few examples of trends in this area in the second part of this chapter.

First of all, one has to give corn its due. Corn is the most widely produced grain in the United States, accounting for more than 90 percent of total value and production of feed grains. Around 80 million acres of land are planted with corn, and while most of the crop is used as livestock feed, it is also processed into a multitude of food and industrial products including starch, sweeteners, corn oil, beverage and industrial alcohol, and fuel ethanol. Without instructions on growing corn from the local natives, European settlers in the New World might not have been successful at their early agricultural efforts. They might have been wiped out entirely, and the

North American continent might be a different place today. Without sweet corn, Americans wouldn't have the signature accompaniment to the backyard hamburger, corn on the cob. From *Oklahoma!* to *Field of Dreams*, and in dozens of other movies, corn appears as the iconic crop. And what movie experience is complete without a bucket of popcorn?

But has corn gone too far? Famed food writer Michael Pollan has made corn, and specifically the overwhelmingly popular high-fructose corn syrup, the poster child for an industrialized food system gone badly wrong.[9] And just because corn is most often grown on farms watered by rain, don't think that water isn't at the heart of figuring out the future of corn.

Right now, an increasing percentage of corn in the United States is going into the production of ethanol. Is this a good thing or a bad thing? The answer here is hopelessly complicated—bound up as it is with issues of water, food, energy, presidential politics, and even (literally) the price of rice in China. Detractors claim that it uses more energy to produce ethanol than a consumer gets out of it, while supporters say that it is environmentally friendly because it essentially turns sunshine into energy that can be burned in a car. The facts haven't changed in the past twenty years, but the politics have, and that has enabled the manufacturing of ethanol to increase over the past ten years. The US Congress passed and President George W. Bush signed a law in 2007 that requires ethanol production to increase from about 7.5 billion gallons to 36 billion gallons annually by the year 2022.[10]

The process of creating ethanol is essentially the same as the process for creating alcohol. Indeed, when it's shipped, producers add gasoline to it so that nobody will steal it to drink and so that they won't have to pay the taxes required on the shipment of alcohol. To make this potent liquid, kernels of corn are flooded with water to create what's called a *mash*. That mash is then mixed with yeast and various enzymes, and after a while, the water is converted into ethanol. What's left is grain that has no starch but otherwise works fine as cattle feed, as we mentioned earlier in this chapter.

Creating that mash is relatively simple. In the factory, all that's needed is corn kernels, the yeast or enzymes, and lots of water. Of course, a lot of water is needed to grow the corn, but in addition, about 4 gallons of water are needed for every gallon of ethanol produced in the distillation and manufacturing process. Some ethanol plants have already overpumped local aquifers to the extent that they have been ordered to shut down, because neighbors successfully sued to get the plant closed. The water costs of ethanol have brought the industry a lot of negative publicity recently.

So maybe all the production of ethanol should be moved to areas where plentiful freshwater already exists? Maybe the Great Lakes? It's true that most of the freshwater in the United States is found in the Great Lakes, but much of the water used in the distillation process is boiled off, meaning that it won't return directly to the lakes the way it might with some other manufacturing processes. Even though the Great Lakes have plenty of water, many people who live or work near the lakes will tell you that they are worried about water levels going down, and they're not easily going to give up their billions of gallons of water. All of the states in the region as well as the neighboring Canadian provinces recently agreed to review the 1909 Great Lakes Treaty, evidence of a strong regional concern about other areas of the country trying to steal their water.[11] Other states that get plenty of rainfall could provide the water but would likely have to build massive reservoirs or underground storage facilities for it, driving up the costs of production even higher.

One technological change may be on the way. The concept is known as *cellulosic ethanol,* so named because it breaks down just about any plant product at the cellular level to release the sugars. Traditional ethanol is made just from the corn grain we eat; cellulosic ethanol is made from corn stalks, cereal straws, and even sawdust and other materials that we generally classify as waste. The material is broken down either using acid or enzymes, and then the regular distillation process kicks in. That means that while cellulosic production is terrific at reducing the amount of water needed to grow the plants, it doesn't improve on the amount of water used in the actual production of the ethanol—again about 4 gallons for every gallon of fuel.

And, of course, longer-term problems are associated with using just the by-product or waste material of the corn-growing process. Right now, the useful part of corn for consumption by people, livestock, or cars all comes from the kernels. The stalks, leaves, and rest of the plant get plowed right back into the soil. Charles Gipp, director of the Soil Conservation Division of Iowa's Department of Agriculture, says that's an important part of farmland preservation and of protecting the environment. "We have serious concerns about anything that takes that material out of the soil," Gipp said in an interview.[12] Tilling in the corn stalks makes the soil much more absorbent. "One of the most important things we can do to avoid flooding and erosion is to ensure that the soil can soak in as much rainwater as possible."

The future of corn, Gipp and others say, is not to be found in massive technological breakthroughs or radically new and genetically engineered plants. The amount of corn grown in Iowa, for example, hasn't really changed all that much over the past twenty years; it's just that the uses have changed, from food for people or cattle, to sweeteners, syrups, and ethanol. Change will happen through small-scale adaptations, such as the one Gipp is managing in Iowa. He is helping farmers construct small wetland areas on their farms. The runoff from their fields can percolate through those wetlands, removing most of the nitrogen and other extra fertilizer components in the process, rather than dumping them into the rivers. Iowa farmers want to do their part to help the environment, and they know that nitrogen in the water is causing massive problems downstream as far as the Gulf of Mexico. That's where algae blooms fed by that nitrogen choke out all other forms of sea life. This is the kind of trend that will continue around the world: farmers doing small, local things that will help not only themselves but also the global water picture. Sometimes this will happen because of government intervention, sometimes because it will save money, and sometimes because it's just the right thing to do.

Any conversation about global water resources always comes back to the fact that 70 to 80 percent of all water consumption on the planet goes into agriculture: to watering the plants and animals that we grow for food. And typically, this water is applied in very inefficient ways, like flood irrigation, where much of the water is wasted. One accelerating trend that is helping make all types of agriculture more efficient and that will definitely explode in the future is the general concept of *more crop per drop*: producing more food from each drop of water used. (And as we'll see, as water becomes scarcer, scarcity concerns may not only force more crop per drop but may also force changes in the type of crop itself: more profit per drop.)

Innovations in irrigation have made it possible for farms around the world to get far better production by using such technologies as (1) center-pivot irrigation systems, which create those big green circles that you see when you fly over the otherwise arid Western United States; (2) drip-irrigation tubes, which slowly drip water into the soil, allowing more efficient utilization and preventing evaporation and runoff; (3) more precise leveling of fields, so that natural or irrigated waters are more evenly distributed; (4) in-place moisture sensors, which determine on a very fine scale where additional water is needed or not needed; and many other innovations. These kinds of technologies can dramatically improve the water efficiency of farms, but the capital costs of such investments can also be quite high. When water is essentially free, incentives are few for the farmer to make any changes. When water starts to get more expensive, these types of investments will be made routinely.

Around the world, the importance of small-scale drip-irrigation and other related technologies can not be overstated. If agriculture could cut its freshwater use by just a few percentage points, this would free up a vast amount of water for other uses. (This of course assumes there is some way to move extra water around, but we'll address this concern in chapter 7.) The shocking thing right now about these types of water saving technologies is how little they are actually used. In places where the dollar value of crops is high and the water costs are growing, like in the Central Valley of California or in Israel, drip-irrigation techniques are growing fast. In fact, innovations in agricultural irrigation are the primary reason that the total amount of freshwater used by the country is actually going down slightly on a per capita basis. Our population is growing, so the amount of water being used isn't dropping much, but it's at least a sign of progress that simple conservation efforts, especially in agriculture, are yielding results.

However, these technologies are catching on frustratingly slowly, even though in many cases they can represent a real lifesaver for the neighbors of the farmer. Government programs to introduce these small-scale technologies have shown some success, and the work of nonprofit or nongovernmental organizations has been profound. The economic drivers of microscale lending and simple small business development have also spread new water reduction technologies around the world in a sustainable way. As water becomes more valuable, and the markets for food and agricultural technologies mature, watch for these kinds of low-tech improvements to become a much more significant global phenomenon.

Many ways exist to use water more efficiently to grow crops. But what about the future of changing the crop itself so that it uses less water? What is the future of the more high-tech changes such as genetic modifications of plants? This is a hot-button political issue, given the hopes and wishes of either the corporations who stand to profit from the marketing of genetically modified organisms or the citizens' groups that are fighting against them. That said, some trends have become well established and will certainly play a part in the future of water use in agriculture.

Advocates of genetic modification touted the ability of transgenic plants to save water when the technology was first gaining traction in the 1990s, but that promise has so far remained largely unfulfilled. Today, advocates for *biotechnology*—as industry advocates prefer to call it—are still saying that plants that can do better with less water because of genetic modifications are at least ten years away. According to the journal *Nature,* scientists have only recently found the genes that control the stomata of some plants, the pores that allow water to evaporate from plants.[13]

Even in areas chronically short of water, the only genetically modified seeds that have been propagated to any significant degree have been those created to tackle pests or weeds. The best sellers here, by far, are plants grown from seeds that have an ability to continue to grow even if doused with a specific brand of herbicide; in other words, the farmer can spray the weeds that spring up in his fields with chemical herbicides without simultaneously killing his money crop. Monsanto produces the most common brand here.[14] As much as half the corn and about 85 percent of the soybeans planted in the United States grow from seeds that have been genetically modified in that way. The percentages are lower in most other countries, but there is almost no country where these types of seeds don't have some share of the cropland.

The other common type of genetic modification is designed to prevent problems with pests. This modification has been especially popular for cotton as a way of fighting various forms of bollworm. The modification to the cottonseed's makeup is to actually make a pesticide a part of the plant's genetic makeup. Because cotton is worn, not eaten, this development has seen less in the way of a backlash from environmentalists. An effort to genetically weave pesticide into potatoes did not go over as well, but the efforts of Monsanto and other companies continue to yield results around the world.

Because drought tolerance has not yet really been bred into any plants used commercially, the actual impact of genetic modifications on freshwater issues is not yet significant. Advocates of genetic modification like to point out that there is a significant positive impact on the water runoff from fields planted with the new seeds because of the reduction in the amount of pesticides used and therefore the amount that is washed off the field and into the surface waterways. This is a significant issue: pesticide runoff is a major contaminant in watersheds, and it's difficult to remove the pesticides in drinking water treatment. This is particularly an issue in developing nations where pesticide use isn't as well monitored. The best estimates, for example, in India are that perhaps 5 percent of the land is planted with cotton; the cotton was responsible for more than half the total pesticides applied before the introduction of genetically modified cotton. Now the sheer tonnage of pesticides used in India may be dropping because of the genetically modified cottonseeds.

THE CASE AGAINST COTTON

Some environmentalists ask why cotton is produced in such wide swaths of the countryside, especially in poor or developing countries.

It's an extremely thirsty crop, using on average nearly twice the water needed for wheat or corn, and it obviously can't be used for food (except when the farmer sells it for cash used to buy food). That economic model for the farmer works when water is free or cheap, as is the case now. If the farmer were to pay actual costs or replacement costs for the water sucked out of the aquifer, the economic model could change dramatically. In a world where Goodwill Industries International and others who salvage clothes from the United States regularly ship massive bales of used clothing overseas, environmentalists say, the world would be better off if those farmers planted food crops that could feed a hungry population.

This leads us to a quick diversion into the water implications of another basic human need: clothing. People need water to drink, water for food to eat, and water to make our clothes. Right now, even the poorest of the poor have something to wear, even if it might be a bit tattered. If water prices climb and cotton production starts to slow down, what will people use for clothing? Remember that washing machine in chapter 3 that only washed clothing made from nanofabrics? Here's where that may come in. Companies are racing to develop textiles with modifications made at the nano, or atomic, level. This technology is new, and emerging slowly, but companies exploring the nanotextiles say they could produce fabric that could essentially repel all smells or that could have the soft, pliable feel of cotton—even when the raw material is made from recycled plastic containers. And then the clothes could be washed with just pulses of air and an ionization charge. This technology seems far-fetched now, but polyester also seemed far-fetched not long ago; later it became fashionable for the rich, and now it's often polyester fabrics that clothe the world's poorest people. As this book is written, it's possible to buy "Isotope Nano Rain Pants made from fully waterproof breathable 2L Nanolite fabric." How long will it be before we see those nanoproducts showing up in the streets of the world's slums?

It's also currently hard to imagine that all-natural cotton could have a bad name among those who consider themselves environmentalists; jeans and a T-shirt are the typical wardrobe of many of those who want to save the planet. Polyester is seen as a symbol for what's wrong with the world. That attitude may have to get turned upside down. As we've said, cotton is an exceedingly thirsty crop. The amount varies based on soil, climate, and other factors, but it typically requires that a field be flooded with about 40 inches of water per season, twice as much as is needed for corn. Although consumer awareness is lower regarding clothing than food, pesticide use for cotton is typically very high, and the runoff from cotton farms is blamed for significant damage to watersheds.

Cotton is almost single-handedly responsible for one of the largest ecological disasters on the planet: the destruction of the Aral Sea in the middle of the former Soviet Union, now Central Asia. At one time it was the fourth largest lake in the world. Every year for a century, it produced as much as 50,000 tons of fish and employed as many as sixty thousand people in and around its seaside cities. The lake is now perhaps 10 percent of its former size, and there is no fishing at all. Those seaside cities are now ghost towns, in view of fishing boats lying on parched ground 60 miles from the closest water. The surface streams that once fed the lake were diverted into the desert where Soviet planners thought they could be better used to produce crops. The biggest cash crop was and still is cotton.

Some efforts are now under way to try to restore the lake, but many experts have essentially written it off, clogged as it is with salt, pesticide residue, and biological weapons developed by the former Soviet Union. International treaties between the individual countries that now share the lake can only be considered a remote possibility. Most significantly, there is still an international demand for cotton, a crop that can only grow with the massive inputs of water from the rivers that once fed the Aral. That trend is expected to continue, and so the Aral will probably be a dead wasteland for decades. If cotton production should ever wane, it might be too late to fix the problem.

In other areas of the world where food is scarce, as in India or much of Africa, it's hard to imagine that cotton is being grown instead of food. It is being grown, however, often because the product is not as perishable. A tomato grown in India can't be sold to a European, but cotton can. Unless some kind of anti-cotton consumer awareness campaign develops alongside nanocoated artificial fabrics that are palatable to the public, look for that harsh economic reality to continue, with farmers growing cotton with water that will not be flowing into the Aral Sea, or water that will not be used to grow food in India or Africa.

MORE ON THE WATER FOOTPRINT

Let's revisit the *water footprint* concept. It's a handy offshoot of the increasingly popular *carbon footprint* used to help people understand and evaluate their fossil energy consumption. It's also a good yardstick to help consumers make wiser decisions about the water implications of their purchases—to measure how much water it took to make any individual product or commodity. In sum, things made out of cotton have a big water footprint, and the bottom-line lesson is this: you might try to save water by turning off the water faucet while you're brushing your teeth, but you'd be saving a lot more water— from a global perspective—by not buying that next cotton T-shirt.

One currently popular line of thinking is that if a consumer wants to be friendly to the environment, purchases should be made locally whenever possible. If you purchase a chair made in China, you are responsible for burning some amount of fossil fuel to get that chair transported from Guangzhou to Galveston. The costs of that fuel, of course, are offset by the price of labor and natural materials in China. (Some evidence exists that those costs are kept artificially low, but for the foreseeable future, the economic policies of the Chinese government, and the desire by the developed world to take advantage of those low costs, show no sign of change.) Labor, energy, and materials costs have always been considered the primary inputs to industrial manufacturing.

But what about the water footprint of that chair? As water prices rise, water will increasingly be thought of as one of the key inputs, or factors of production. And the question will increasingly be asked: is there a conflict between the water footprint and the carbon footprint? In the case of food, the answer is not always clear cut. Remember, agriculture primarily does the job of converting water into food. So, those who believe that only locally grown food is environmentally sound are forgetting one thing: it may be that the water itself was transported great distances by means of nonrenewable energy resources. In other words, a low carbon footprint may lead to a high water footprint, or vice versa. As water becomes more expensive, pressures will increase and environmental sense will require that we grow more of the world's food in areas where water is plentiful, rather than just outside of town.

Consider the Imperial Valley of California. It's possible to find oranges, tomatoes, and apples at farmers' markets all over the greater Los Angeles and San Diego area every day of the year that have been grown in that valley. Many shoppers purchase the goods there, thinking they are doing a good thing for the local farmers and the environment. Certainly they are helping the local farmers who can sell directly to the consumer and typically make a much better living than they can by selling to wholesalers. They are also helping the environment in the sense that the goods purchased come in trucks from just a hundred or so miles away, and not the 2,500 miles an orange might have to come from Florida.

From the water perspective, however, the picture is quite different. The Imperial Valley was christened by boosters around 1900 because they thought the name the locals called it, "the valley of death," was a bit too negative. The developers of the Los Angeles megalopolis knew that if they could divert the Colorado River into that valley, they could use the fact that it's always hot there to grow two or more season's worth of crops every year to feed the burgeoning population of Southern California. The first farms were located in the lowest areas of the desert, and the water flowed in so abundantly from the Colorado that it drowned them in the desert and a new lake was formed. They didn't come up with a PR-friendly name for that lake; it was so salty that it became known as the Salton Sea, and it's still called that today. The farmers relocated from the area of what is now the Salton, and all the runoff from the alfalfa, citrus, and other cropland continues to drain in. The Salton Sea is saltier than the Pacific and is now the largest lake in California.

So, what's the problem? Well, according to the local farmers, there's no problem at all. The problem exists, however, for those downstream in Mexico who once farmed what had been one of the most verdant areas in North America. They aren't farming there anymore, and also gone are the people of the Cocopah tribe, the descendants of whom are now operating a casino in Arizona. The US government paid for a $300 million desalination plant to treat the water just before it returned to Mexico, but for reasons of politics and economics that could make the most hardened cynic blush, the plant has never been turned on.

Growing alfalfa or vegetables in the Imperial Valley requires extensive irrigation with water brought from far away. The only way an orange tree can produce fruit in the Imperial Valley is with water taken from the Colorado River and poured onto what would otherwise be a rock-hard desert. So if a consumer in Southern California really wanted to be environmentally careful from a water perspective, perhaps the best thing that person could do is insist on purchasing only oranges grown in Florida, where drought-level precipitation is 30 inches per year, ten times the average annual precipitation of the Imperial Valley. In other words, maybe the negative implications of the water footprint here outweigh the negative implications of the carbon footprint for the same delivered orange. As water becomes scarcer and scarcer, society will increasingly face these kinds of tough trade-offs.

SUMMING UP WATER AND FOOD

It's not really the place of this book to prescribe what consumers *should* be doing, but we do want to try to predict what they *will* be doing in coming years, and what the consequences could be. Environmental concerns have had a significant impact on consumer purchasing decisions over the past thirty years, and especially the past ten years. This is an increasingly pertinent consideration in the minds of many consumers, and it will not be going away anytime soon, if ever. Largely absent from the conversation about environmental issues to date, however, has been the question of water. This will change, as more and more stress is placed on the water system. Politicians, environmental leaders, and citizens will increasingly recognize the critical role that water plays in these issues, and the cost of water will increasingly become a factor in the minds of consumers, just as the impact of carbon-based fuels has become more of a purchasing factor already. The term *fossil fuels* is common now; less common is the term *fossil waters,* referring to waters withdrawn from an aquifer that essentially can't be replaced. We'll be discussing this more in chapter 6, but in terms of food, don't be surprised if you see on packaging in the coming years, "Irrigated with natural rainfall, no fossil waters used."

Also, look for growth in what is now a tiny niche but one that is expanding fast: community-based urban farms. They are already starting to pop up in cities around the country, and indeed, in many cities around the world. *Urban agriculture* is an increasingly critical source of food and sustenance to more and more people, particularly in less developed regions. Urban agriculture represents a real paradigm shift in food production, using both intensive food production techniques as well as improved urban waste treatment approaches. It means growing food closer to where it is needed, both to reduce transport costs and to enhance freshness. It also means taking urban waste streams and doing something more productive with them, so that the nutrients in that "waste" stream can be used for food production and for good instead of literally being sent downstream where they cause problems for others.

So, where are we going with the future of water in agriculture? There is a serious dichotomy here in what experts predict. On the one hand, we have the environmentalists who predict massive changes, with the aquifers drying up, major dam projects becoming extinct, and a growing population taxing natural resources to the breaking point. On the other hand are the agricultural producers who see changes coming, but who continue to utilize new technologies and innovative practices to continually adapt and meet the needs of a growing population. Which side will prevail?

As always, the truth probably lies somewhere in the middle. Agriculture will certainly see some very dramatic changes in the coming decades, just as it has in the past. Perhaps the most famous example is the so-called Green Revolution of the 1960s—made possible by genetically enhanced grains (particularly rice), new fertilizers and pesticides, and large-scale irrigation—which helped avert famine in India. At the time, this was viewed as a technological success, a way to feed the ever-expanding population of the earth, a way to defeat the Malthusian predictions of pessimists. But as we know now, those practices also encouraged and eventually led to overuse of water, the flushing of important nutrients from the soil, and the contamination of groundwater in many places, in addition to a plethora of political, economic, and social questions that are still unanswered. From today's perspective, many would argue that the Green Revolution simply delayed the day of reckoning.

In the future, either as a result of new government policies or the simple reality of aquifers running dry, we're likely to see many large irrigation projects simply drying up, and large tracts of currently productive agricultural land becoming too salty to use. (We refer you to Sandra Postel's excellent, if dire, books on this topic, *Last Oasis* and *Pillar of Sand*.[15]) These trends will all be converging in the next fifty years. When that happens—whether gradually with one or a series of droughts, dam failures or decommissioning, or other forced changes or disasters—it may come to pass that we no longer grow citrus fruits in the deserts of California, just as we no longer grow cotton around the Aral Sea.

Agriculture, however, is flexible. Each season the farmer has to decide what crop to put in the ground, which fields to use, and what kind of animals to raise. Today, we are busy shifting water from agriculture to (currently) more valuable urban and industrial uses. In the future, however, if and when too many farmers start to run out of water, or when the population starts to get hungrier, we may decide to reverse that and shift more water usage back to agriculture again. As markets, tariffs, transportation costs, and alternative water sources adapt and change, farmers will also adapt and make sure that they continue to produce food with every drop of water that they can find.

❧

1   Gary C. Smith, "The Future of the Beef Industry." *Proc., The Range Beef Cow Symposium XIX, December 6–8, 2005, Rapid City, S.D.* http://beef.unl.edu/beefreports/symp-2005-02-XIX.pdf.

2   Water Footprint Network. "Product Gallery: Beef." http://www.waterfootprint.org/?page=files/productgallery&product=beef.

3   J. L. Beckett and J. W. Oltjen, "Estimation of the Water Requirement for Beef Production in the United States". *Journal of Animal Science* 71(4): 818. http://jas.fass.org/cgi/content/abstract/71/4/818.

4   Interview with Foster Stroup by Scott Yates, April 2010. See also http://www.noamkelp.com/.

5   Phone interview with Scott Yates, March 2010.

6   Quoted in Brendan Borrell, "Sausage Without the Squeal: Growing Meat Inside a Test Tube." *Scientific American.* March 31, 2009. http://www.scientificamerican.com/article.cfm?id=test-tube-pork.

7   David Pilz, "Charting the Colorado Plateau Revisited." Sustainable Development Workshop, Economics Department, Colorado College. http://www.coloradocollege.edu/Dept/EC/Faculty/Hecox/CPWebpage/issuespageTurner.htm.

8   Juliet Eilperin, "World's Fish Supply Running Out, Researchers Warn - washingtonpost.com." http://www.washingtonpost.com/wp-dyn/content/article/2006/11/02/AR2006110200913.html.

9   Michael Pollan, *The Omnivore's Dilemma: A Natural History of Four Meals.* Penguin Press, 2006.

10 US Congressional Research Service. "Energy Independence and Security Act of 2007: A Summary of Major Provisions." http://energy.senate.gov/public/_files/RL342941.pdf.

11 See for example *Economist*, "The Great Lakes' Water: Liquid Gold," May 20, 2010, http://www.economist.com/node/16167886; *Federal Statute on Great Lakes Water Diversions*, Ohio Department of Natural Resources, http://ohiodnr.com/water/planning/greatlksgov/fedstatut/tabid/4053/Default.aspx; or "Great Lakes Diversion and Water Use," *Stop Asian Carp: Protect our Great Lakes. A Project of Michigan Attorney General Mike Cox* at http://www.stopasiancarp.com/agcox.html for a discussion of concerns about diversion of Great Lake water outside of the Great Lakes Basin.

12 Phone interview with Scott Yates, March 2010.

13 Shigeo S. Sugano et al. "Stomagen Positively Regulates Stomatal Density in *Arabidopsis*." Nature 463, 241–244. January 14 2010. http://www.nature.com/nature/journal/v463/n7278/full/nature08682.html; see also Barbara Miller, "Breakthrough Could Lead to Drought-Resistant Plants." *Australian*

Broadcasting Corporation News, Feb. 28, 2008. http://www.abc.net.au/news/stories/2008/02/28/2175160.htm.

[14] *Economist*, "Monsanto: Lord of the Seeds." Jan. 27, 2005. http://www.economist.com/node/3600040.

[15] Sandra Postel, *Last Oasis: Facing Water Scarcity*. W. W. Norton & Co., 1997; *Pillar of Sand: Can the Irrigation Miracle Last?* W. W. Norton & Co., 1999.

CHAPTER 5

THE FUTURE OF INDUSTRIAL WATER USE

The industrial consumption of water is one of the more misunderstood uses of water around the world. Most people are familiar with water use in the home because that's where they live and use water. Water use in agriculture is well studied, tracked, and accounted for simply because farms are so visible and because everyone has to eat.

Everyone eats food and drinks water, but we also all consume a lot of water, albeit indirectly, in the production and manufacturing of all the other goods and products that we use every day—things as big as an airplane or as small as a paperclip. The one ingredient that makes it all possible is water. The populace understands less about water use in this area, even though water is at the heart of nearly all manufacturing plants.

That has always been the case. The earliest factories couldn't have existed without power from waterwheels. The advent of steam power changed the equation some, but not much; fossil fuels were originally used simply to boil water, and the explosive power of steam made possible the factories and trains of the industrial revolution. Water was also used to cool and clean all the factories. Even now, with the ability to move water great distances, new industrial zones most often crop up in areas where water is plentiful. The fastest-growing cities in the world today are mostly the industrially oriented ones, like Shenzhen in southern China. That city has grown from a fishing town of about seventy thousand people in 1980 to what it is now: a giant manufacturing city that has as many as twenty-two million people, nearly all of them working to produce so much of what people buy around the rest of the world. Shenzhen has been able to sustain all of that growth because it lies in a water-rich delta, where the factories can use water essentially for free.

So, what's the future of water use in industry? This sector may be the hardest to predict with any accuracy. With water use around the home, for instance, it would have been relatively easy and accurate to predict fifty years ago that most water use would go to irrigate lawns around the home over the next fifty years. Today, it's pretty easy to predict that landscaping around homes in the future will feature far less grass than we see today.

It's easier to understand the problem of predicting industrial water usage if we think back and realize how hard it would have been years ago to predict where we are now. In 1960, most countries made the things that were consumed within their own boundaries; international trade was modest by today's standards. An imported toy from Germany was rare and expensive. Today, toy stores the size of supermarkets are filled with toys made mostly in China. In 1960, an imported cigar from Cuba was common enough, and French or Italian wine was pricey but well known. Today, the ban on Cuban cigars is nearly fifty years old, but wines from the United States, Europe, South America, Africa, and Australia all jostle for shelf space in liquor stores. Nobody was predicting fifty years ago that a sleepy fishing village in southeastern China would become the factory to the world that it is today. So, predicting the next fifty years of the industrial use of water is more than a little difficult.

EMERGING TRENDS

That said, some emerging trends here may portend the future. First is that industry will increasingly tend to locate where water is plentiful, reversing a trend over the past fifty years that has seen water largely being moved to industry. Major industrial operations have basically ignored the cost of water when locating manufacturing facilities and focused instead on finding cheaper labor or cheaper energy sources. As water prices rise, industrial installations that use a lot of water will, naturally enough, start to be located in areas of more plentiful water.

The second trend is that water will be used much more efficiently within the factory fence—to conserve water and reduce usage, to recycle wastewaters for reuse in manufacturing processes, and even to prevent wastewater discharges altogether in what's called *zero liquid discharge*. More efficient industrial water usage is growing rapidly around the world, because it makes environmental and economic sense, and it will accelerate even further as water prices rise.

Finally, the third big trend is that new and emerging technologies may allow us to dramatically reduce the overall need for water in the first place, through improved processes or through the reuse of water many times over.

As all of these trends start to gather steam and interact, we will gradually cease to make such a distinction, as we do today, between wastewater and water. Instead, we will start to think of wastewater as just one more source of water, and the boundaries between water and wastewater will start to fade. We'll look at each of these major trends in more detail below.

WHERE THE WATER IS

In 1900, dozens of breweries thrived in Milwaukee, Wis. A combination of good access to grains, lots of German immigrants, and seemingly unlimited water from Lake Michigan gave brewers all the resources they needed to produce as much beer as the country could drink. A vibrant tanning industry also existed then, processing the hides of animals into leather for use in many of the other factories that surrounded the region. Water is the chief ingredient in beer, and the process of brewing beer requires a lot more water. Tanning also uses lots of water, most of which gets fouled with dangerous chemicals.

By 2000, most of the breweries and tanneries and much of the rest of the industrial base of Milwaukee had pulled up and left. Most leather is now processed in China, and domestic beers are more commonly brewed in the South. Cheaper labor is the reason in both cases, though lax environmental regulations in China certainly are a part of the equation. Over the past several decades, the availability of water has not really been a factor in terms of where businesses locate their plants and factories. The Atlanta area, for example, pulled as much water as it needed from Lake Lanier while various industries set up shop and put a growing (and largely nonunion) population to work.

Fast forward to the present, and take another look at these two metro areas to see how water availability is now trumping both labor and energy in terms of where industries choose to locate. US District Judge Paul Magnuson ruled in 2009 that Atlanta's removal of water from Lake Lanier is illegal, and gave the city, the state and the US Congress three years to figure out what to do about it. As of this writing, it's not exactly clear what will happen, but the options all involve spending plenty of money on new water rights, new infrastructure, and more. Undoubtedly, water intensive industries are wondering if they'll be able to afford to continue operating in Atlanta. And what is beginning to happen in Atlanta now could happen in other cities in the future.

At least that's what some in Milwaukee are hoping. A businessman with deep roots in that Wisconsin city, Richard Meeusen, chief executive of Badger Meter, a major manufacturer of water meters, has been spearheading a plan to attract new business to Milwaukee with the promise of cheap or even free water. "I've got to think that there are plenty of wet industries down there [in Atlanta] worried about what's going to happen," Meeusen says.[1] It's certainly possible that cities like Milwaukee could attract a large influx of new water intensive businesses, given that the municipal water provider in Milwaukee pumped about 54 billion gallons of water in 1970 and 34 billion gallons in 2009, even though the service area has grown substantially over the same time period.

In other words, in the future, high-growth industries may start to seek out cities like Milwaukee simply because of the availability of inexpensive and abundant water. Milwaukee certainly hopes that is going to be the case, and city leaders have declared it their mission to become the "world's water hub for research, economic development, and education" through a new Water Council and the growth of the Great Lakes Water Institute—a part of the University of Wisconsin at Milwaukee.[2] That trend has already begun, and the city has already seen growth in a new industry scarcely imagined just a generation ago: data centers, sometimes called *server farms*. These are the computers that power the Internet, and while they have gotten more efficient and faster every year, they still take a lot of power to run, and consequently they always generate a lot of heat—heat that must be cooled with water.

Availability of abundant water is a big reason that Oregon has also done so well in attracting data centers. A generally steady volume of naturally occurring surface water available in that state powers the hydroelectric facilities common in the region. That same water can be put to use in new forms of evaporative systems to cool down the thousands of computing devices inside the data center. Facebook, Google, and other big names in the Internet world are all building massive data centers in Oregon, many of them on the banks of the Columbia River.

REDUCING CONSUMPTION

Other users of water are the manufacturers of the chips that go into those computer servers and into the laptops, video games, and countless other electronic devices that permeate modern life. While these chip makers are sensitive about revealing their trade secrets, the process is essentially one of taking sand and turning it into a semiconductor (something that can conduct electricity under certain conditions and not conduct it under other conditions). That metal on a semiconductor chip is etched with the smallest of lines. After the etching, water is used to wash away the bits that have been carved out of the chip. By some estimates a chip may need to be washed four hundred times in the process of manufacturing. That's a lot of washing; it adds up to a lot of water, and the water must be ultrapure, clean of almost any chemical constituents and in fact much cleaner than the water we humans routinely drink.

Intel Corporation is a good example of the second trend we mentioned: the company is redesigning its manufacturing process and instituting recycling processes so as to consume less water. The largest of the chip manufacturers, Intel uses billions of gallons to make its semiconductor chips. Being such a high-profile manufacturer, and because so many of its facilities are in the United States, Intel goes to great lengths to tout its water friendly ways. The company says that as of 2010 it reclaimed 3 billion gallons of water a year.[3] In other words, the chip maker now diverts and reuses 3 billion gallons of water that had previously been sent to the nearest municipal wastewater treatment facility and instead uses it for cooling or other water needs.

Still, making those chips uses a lot of water, and in part it's because of this requirement to utilize ultrapure water. Not only must the water not have any particles that could damage the chip during the sensitive manufacturing process, but it also must be so pure that it can't really conduct electricity. Completely pure water doesn't conduct electricity; it's all the other particles suspended in lake and river waters—and even good clean drinking water out of the tap—that allow an electrical current to pass through. This is a big deal from the perspective of a chip manufacturer because the process of creating 1 gallon of ultrapure water creates 2 more gallons of water that can't be used in the manufacturing process.

And, in an illustration of our third major trend, Intel scientists are working hard on new technologies and processes that will minimize the number of times a chip needs to be washed, and on various other improvements to the currently water intensive ways of doing things. Some research institutions and more theoretical scientists are using nanotechnology or other nascent technologies to make chips in radically new ways. In fifty years, some new technique will probably be common, but for the near-term future, the manufacture of chips will likely continue to use significant amounts of water.

And from the perspective of people in these high-tech industries, there's nothing wrong with that: they would argue that the water they're using is creating far more value to the economy than many alternative uses. For example, by one calculation, the amount of water that it takes to produce $100 worth of alfalfa can produce $5 million worth of computer chips. That's enough revenue to provide dozens of jobs, and it may even leave a little left over for more local watershed protection, environmental restoration projects, and research into further water conservation measures. Intel in 2008 pledged to spend $100 million to do just that, and has an entire division devoted to reducing water use.

The manufacturing of computer chips gets a lot of attention, but in the larger industry picture, it is a relatively small player. Intel reports that it uses approximately 8 billion gallons of water per year at its factories around the world.[4] In contrast, International Paper reported that it used 211 billion gallons per year in 2008.[5] Like Intel, International Paper is touting its efforts to reduce the amount of water used in its plants, and the amount of wastewater discharged, by reusing more water, especially the water that is only used for cooling. The paper giant says that it has reduced the amount of wastewater flowing out of its plants from 194 billion gallons in 2004 to 187 billion gallons in 2008, even though the company opened a new factory during that period. Put another way, the *reduction* in yearly effluent from International Paper represented nearly the same volume as *all* the water used by Intel in a year.

Different industries need different amounts of water to produce their products. For example, the environmental think tank Sustain-Ability listed reform of the paper industry as second only to agriculture in terms of needing major improvements in water use habits over the next ten years.[6] Just as we mentioned in the case of agriculture, even a small percentage savings in the amount of freshwater used by industry, particularly the more water intensive sectors, can result in a huge amount of water that can be used either for other purposes such as drinking water or simply left in place to help maintain the water table or the quantity of water in the lakes and rivers around the factories.

One of the biggest uses of water in many factories, and particularly in the thermal power industry, is for cooling. Water does an excellent job of removing the heat from a newly minted ball bearing, baby bottle, or baseball bat. Often the water used for cooling never even touches the product; it just flows through the machinery and lowers the temperature a couple of degrees, then flows out again. In cars, special additives go into the water used to keep the engine from overheating. Moving automobiles obviously can't use the waters flowing in a nearby river for cooling their engines, but factories and power plants can. And because that water from the river is essentially free, we see many of these most water intensive industries—power plants, petroleum refineries, or pulp and paper manufacturing operations—located along major waterways.

If the price of water keeps climbing, more factories will probably follow the lead of Intel and others in figuring out ways to use the same water two or three times—or maybe hundreds of times: reusing wastewater or storing it for future use rather than just discharging it back into a river or sending chemically contaminated waters to the local municipal wastewater treatment facility. Many of the future solutions here are not that fancy or high tech and may include such measures as simply building large water storage facilities on the site of the factory. If the only thing the water is doing is lowering the temperature of a machine by a few degrees, the same water could be used over and over, particularly if it is stored underground where it can gradually cool off. Storing water isn't hard or complicated, but it can cost considerable capital investment dollars. Today, storage may only make sense for factories when the price of extracting water from underground aquifers or getting it from a municipality begins to cost more. In the future, with rising water prices, these types of simple solutions will become much more common.

LONGER-TERM TRENDS

Over time, industrial users will increasingly realize that all these different "types" of water are *still simply water*. Various industrial or municipal wastewaters, just like groundwater, surface waters, and even pristine mountain lake water, should all be viewed really as just potential sources of raw water for industries or water utilities to use in the provision of clean water. In the future, water utilities may well compete fiercely over their rights and access to wastewater streams as untapped water sources become more restricted and hard to find.

Indeed, an investment group recently paid $70 million for the rights to a large treated-wastewater flow in Prescott Valley, Ariz., as a means of replenishing an aquifer below a nearby area that was ripe for commercial development.[7] What we call wastewater today will increasingly be thought of as a resource rather than a waste.

Another interesting way to look at water use by industry is to examine it through the lens of how much water is actually consumed. Figure 5-1 on page 90 provides a quick look at the overall usage of water in this country. An important distinction between industrial and agricultural use is the difference between *consumptive* and *nonconsumptive* use of water. Power generation, for example, is a huge industrial user of water, but most of that use is for nonconsumptive cooling purposes. In other words, the vast majority of the water is not actually consumed but just passed through the plant for cooling and then returned to the river. This is just one of a number of examples of the complex but critical factors at the nexus of energy and water.

Let's look at one example: American Electric Power, based in Columbus, Ohio. That company owns eighty power plants, mostly in the Midwest. It has been moving its power plants from *once-through* water cooling systems to *closed* water recycling systems. As of 2009, nearly two-thirds of the power company's steam-generated electricity came from plants using closed systems. How much water does this save? Consider that a typical coal-fired power plant can use up to 400,000 gallons of water per minute to cool its machinery. A closed system that recycles the water over and over actually uses about 1.3 percent of that much water, or about 5,300 gallons per minute. This, in a nutshell, is the difference between consumptive and nonconsumptive use of water. In a once-through plant, as much as 98 percent of the water is drawn into a factory and then immediately discharged back into the surface waterway again.

Even with a closed-system power plant, about 325 million gallons per year of water are still consumed. Even if every single one of the more than five thousand power plants in the United States were to switch to a closed water cooling system, they would still need more than 2 trillion gallons of water per year—and we aren't anywhere close to producing all our power with closed systems. However,

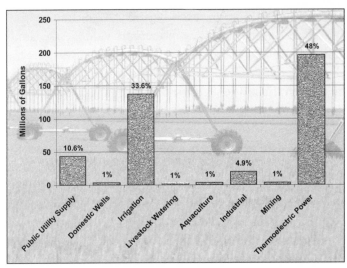

Source: Hutson et al. 2004.

FIGURE 5-1 AMOUNT OF WATER USED IN THE UNITED STATES FOR COMMERCIAL, INDUSTRIAL, AGRICULTURAL, AND DOMESTIC PURPOSES

many power plants along our coasts are once-through systems that use seawater for cooling. Indeed, that's one of the few things that can be done with seawater in our modern society—and explains why 96 percent of all seawater used in US industry is used in cooling thermoelectric plants, according to the US Geological Survey. The USGS also reports that about half of all surface water used in the United States is used for cooling power plants.[8] That number may seem high, given that agriculture consumes about 80 percent of the freshwater used in the United States. However, two factors explain this apparent conundrum. First, thermal cooling is nonconsumptive. Second, much of the water used in agriculture comes from groundwater that is pumped to the surface, not from rivers and streams on the surface.

ENVIRONMENTAL CONCERNS

The use of ocean water for cooling in power plants seems like a simple answer for coastal plants, but it is also not without controversy. Some environmentalists say that raising the temperature of the seawater changes the local environment near the discharge pipes; that's certainly true, but the impact of a few power plants discharging warmed-up waters into the ocean is pretty small and localized. As a society, we have to balance that potential impact with the potential damage that could come from heating up much smaller and more localized freshwater sources like rivers or lakes or depleting underground aquifer resources that are far more precious. In addition, seawater can't be used as efficiently in closed systems because of the increased buildup of salt. So seawater cooling is no panacea, either.

This balancing act that power companies must accomplish is only going to get trickier in the future. A coal-fired power plant in the United States has a vast maze of environmental regulations with which it must comply, mainly from the US Environmental Protection Agency. Clean Air Act regulations require new and better techniques of scrubbing nasty contaminants out of the smoke that comes from burning coal or natural gas. To do that scrubbing, more water is needed, and after it has cleaned that smoke, the remaining sludge is loaded with noxious chemicals. In short, air pollution has been morphed into water pollution. And those sludges can't be just dumped in a river, because that in turn would violate clean water laws. This example only starts to give an idea of the complexities of generating electrical power in an environmentally sustainable manner. While the loudest arguments today against coal-fired power plants come from those concerned about greenhouse gas emissions, in the future we'll likely see more and more concerns about the tremendous amounts of water needed to generate all that power and to keep all the air emissions relatively clean. This is just one more illustration of the delicate interchange between energy and water, one that will continue to grow in importance over the coming decades.

AN EXAMPLE FROM THE COAL INDUSTRY

Just as with agriculture, where we examined the amount of water needed to produce beef, it's helpful to look at the full water footprint of producing electricity, especially electricity produced with coal.

Even the most efficient coal-fired electric plants, as we just saw, need vast amounts of water to operate, but what about the water needed to get the coal out of the ground and to the power plant to begin with? The answer, as you've probably guessed, is a lot. The exact number is hard to pin down, mostly because the USGS only makes specific reports for twenty-two states, those that reportedly have most of the mining activity. The water used in those twenty-two states for all types of mining is 3.5 billion gallons per day, or about 1.25 trillion gallons per year.[9] Put another way, the mines in twenty-two US states use as much water in two days as Intel uses all year in all its factories around the world to produce those computer chips. Of course, Intel also uses mined materials to make its chips and electricity to run its factories, so this usage is also a part of the water picture for Intel—in just the same way that pumped groundwater growing alfalfa for cattle feed is part of the hamburger water footprint.

And the total water footprint of electric power production may extend beyond simply the consumption involved with generating the power and, in this instance, mining the coal. The location of the mining operation and how close it is to the power plant may add even further to the overall water footprint of those kilowatt-hours of power that are eventually delivered to our homes. The coal may have to be hauled hundreds of miles by rail, consuming further water and energy.

Or, take the example of the Black Mesa coal mine in northeastern Arizona, located on the Navajo and Hopi reservations in the high desert of the Southwestern United States. This mine is owned by the largest privately owned coal producer in the United States, Peabody Energy Corporation, based in St. Louis. For years, the mine used at least 1.5 billion gallons of water every year to transport the coal from the mine through a 273-mile slurry pipeline to the Mojave power plant in Laughlin, Nev.

Peabody ground up the coal and mixed it with water to create a slurry that it pumped to Nevada, having figured that this was cheaper than moving the coal by truck or rail. The water to produce this slurry was pumped out of an underground aquifer near the mine, depleting a resource that had been there since the last ice age. The whole scheme worked financially only because the company could pump that groundwater up to the surface and reimburse the tribes $1.67 per acre-foot, or about a half a cent for a thousand gallons. That's about the same cost that many farmers pay for subsidized water. In the case of farmers withdrawing groundwater, however, much of the water used for irrigation at the surface eventually filters back down to help replenish the aquifer. In the Black Mesa case, the water was pumped up out of the aquifer, then sent into another state with no chance of returning. Local environmentalists pointed out that the pumping for the mine dried up springs that had been feeding local farms for nearly a thousand years—some of the longest continuously occupied communities anywhere in North America.

The slurry pipeline operated from 1969 to 2005. At the end of 2005, the US government shut down the pipeline operation. That decision shuttered the mine. Peabody applied to open the mine again, and in the last days of the George W. Bush administration, Peabody received permission, but never got its process up and going before environmental groups sued to stop the mine. Peabody recently lost another court ruling, and now plans for the mine are on hold.[10] If it does reopen, it has applied to the government to use as much as 400 million gallons of water a year for mining operations. That's way down from the 1.5 billion it was using when it slurried the coal, but it's still a lot of water removed from underneath a desert.

Pipelines to move coal appear to be dead, but pipes to move water will continue to be with us. Just how big those pipes get remains to be seen. Certainly, some big ones have been proposed. Take, for example, a proposal from the 1980s that would have seen water shipped from the Great Lakes to Wyoming, and then returned to the Great Lakes region carrying coal. That plan was shot down because the people surrounding the Great Lakes rose up in massive opposition to any plan that would divert water away from their basin. That sentiment remains today, and even one recent proposal that would have used water from Lake Michigan just a few miles outside of the lake drainage area to process corn and turn it into ethanol was quickly shot down. This regional concern prompted by that giant coal sluice, among other projects, ultimately led to a series of water agreements, culminating in the Great Lakes Basin Compact, agreed to by the legislatures and governors of eight states and then passed by the US Congress and signed by President George W. Bush in 2008.[11]

GOLD MINING

The Black Mesa situation may have been unique because of the massive amount of water it used just for the transportation of coal, but all coal mines still use plenty of water, generally in the cleaning and refining of that coal, and as a control so that the flammable coal will not catch fire when it's scraped, drilled, and excavated by heavy mining equipment. In the mining of other rocks, minerals, and gems, water is not typically used so much for the digging and extracting as it is for the processing. To this day, gold mines use water like the forty-niners did, albeit with far more advanced technologies, to separate and wash away the barren rock and concentrate the heavier gold. Because little of the water is actually consumed in that process, the concern here is generally one of wastewater treatment. In the United States and in Europe, mine wastewaters are tightly controlled by governmental regulations. In other parts of the world, the environmental controls typically are more dependent on the goodwill of the mine operator. That goodwill is more significant now than ever, and mining companies often like to tout their environmental good works.

For example, one leading global gold mining company, Newmont Mining Corporation, says if local regulations of the country in which it is operating are less stringent than those of the US Environmental Protection Agency, it will run its operations so as to comply with the US standards.[12] The company says that it also tries to find innovative ways of reclaiming the environment in which it works, and many of them are related to water. For example, the company touts its work in Peru, where it operated an enormous gold-mining pit high in the Andes.[13] Rather than fill it back in, Newmont turned the pit into a reservoir with a capacity of 1.5 billion gallons of water. We'll get into the various aspects of dams in chapter 7, but this situation is unique. Because the pit is not stopping the flow of any river, it doesn't have many of the environmental consequences of dam on a river. This reservoir does provide water for a region that gets voluminous rain part of the year and almost none the rest of the year. It's certainly not something that the local farmers could have afforded on their own, but with the subsidy of the money made by the gold mined during the effective life of the mine, it has proven to be a windfall for local residents. According to Newmont, more than five thousand farmers and ranchers use water from the reservoir.

Certainly, many environmental activists will say that just about any mining is bad for the environment, and it's hard to argue that stripping off the land or leaching chemicals through the dirt to extract minerals is environmentally friendly. On the other hand, modern life, for better or worse, is thoroughly dependent on gold for everything from jewelry to the connectors inside cell phones. We all need a whole range of other mined materials in nearly every aspect of modern life. Other minerals make possible the building blocks of our modern society, from bicycles to jet airplanes, computers, and solar panels. And water is a critical manufacturing input in virtually all of these products.

THE NEXUS OF ENERGY AND WATER

The future of solar cell manufacturing, for example, is huge. Solar cells are made with mined quartz treated with boron and phosphorus, which allow the electricity to flow through mined copper wires. The manufacture is similar to that of computer chips and requires large amounts of purified water that needs two or three gallons of water to produce one gallon of water clean enough to use in the solar cell manufacturing. Once installed, however, a photovoltaic cell eliminates the need to use water intensive coal and coal-fired plants. Here again the delicate balance of energy and water, environmental concerns and consumer demand, all play out.

Other difficult but intriguing energy questions will increasingly perplex planners going forward. For instance, consider this question: How many gallons of water will it take to produce a barrel of oil from the Canadian tar sands? What do we do with the resulting contaminated process water? The massive Marcellus shale natural gas deposits underlying Pennsylvania and New York, currently touted as an avenue to US energy independence, are laden with various water problems that are now slowing and may eventually stop its development. Similar water usage and disposal issues challenge the exploitation of coal bed methane, another source of natural gas. The questions are legitimate, as people near the project wonder where all the necessary water is going to come from. A lot of surface water is needed to pump underground to fracture and collect that gas. How will the natural and injected waters resulting from the process be treated or disposed of? Will the drilling process itself irreversibly contaminate groundwater aquifers that local residents and cities have depended on for hundreds of years?

On the other side of the globe, how many barrels of Saudi oil will it take to produce a gallon of desalinized drinking water? The rapid modernization of the arid Arabian Peninsula rests upon the consumption of a large portion of their natural energy reserves to "manufacture" water for their people. As much as 20 percent of California's total electricity consumption reportedly goes to moving water around the state and treating it.[14] China is spending perhaps $100 billion or more to create huge diversions of water from the south to the north of the country.[15] These projects will require massive amounts of energy into the long-term future to pump the water to the big cities of the north.

Many other interesting challenges and trade-offs are emerging at the intersection of energy and water. Most of these issues can be reduced to this: it takes lots of water to produce energy, and it also takes lots of energy to treat and move water around. The rub is that the demand for one could soon cripple our use of the other. As Michael Webber put it in the October 2008 edition of *Scientific American*, "many people are concerned about the perils of peak oil—running out of cheap oil. A few are voicing concerns about peak water. But almost no one is addressing the tension between the two: water restrictions are hampering solutions for generating more energy, and energy problems, particularly rising prices, are curtailing efforts to supply more clean water.... the situation should be considered a crisis, but the public has not yet grasped the urgency."[16]

So the future of water use in industry clearly mirrors the trends that we saw in agriculture and in the home. In short, it just won't be possible anymore, for instance, for a coal mine in the high desert to extract massive volumes of ice age water from thousands of feet down, use that water to transport coal to a power plant in another state, and then throw the water away. In the future, industry will be forced to think of water not as a free good but as another critical factor of production, just like we think of labor and energy today. Water intensive industries will increasingly have to build factories in areas where water is abundant. That doesn't always happen now, because our current system radically undervalues the water. As the price of water starts to catch up to its real value, industry will have no choice but to consider water more as a precious resource and less as a free commodity.

ဢ

[1] Phone interview with Scott Yates, April 2010.

[2] The Greater Milwaukee Committee, "The Milwaukee 7 Water Council." http://www.gmconline.org/index.php?option=com_content&task=view&id=150&Itemid=76.

[3] Todd Brady, "A Water Policy?" March 17, 2010. *CSR@Intel: Putting Social Responsibility on the Agenda.* http://blogs.intel.com/csr/2010/03/a_water_policy.php.

[4] Intel Corporation, "Water Footprint Analysis." Intel 2009 *Corporate Responsibility Report.* http://www.intel.com/Assets/PDF/Policy/CSR-2009.pdf#page=42.

[5] International Paper, "Global Water Use." 2010. http://www.internationalpaper.com/us/en/company/Sustainability/PF_GeneralContent_1_3601_3601.html.

[6] SustainAbility, "Issue Brief: Water." http://www.sustainability.com/library/issue-brief-water.

[7] See "Arizona Water Rights Auction Tops $20/m^3," *Global Water Intelligence* 8:11, November 2007, page 22, for a description of the purchase of wastewater effluent rights in Prescott Valley, Ariz., by Water Asset Management, Inc. http://www.globalwaterintel.com/archive/8/11/general/arizona-water-rights-auction-tops-20msup3sup.html.

8 Susan S. Hutson et al., *Estimated Use of Water in the United States in 2000.* March 2004. US Geological Survey Circular 1268. http://pubs.usgs.gov/circ/2004/circ1268/htdocs/table03.html.

9 Ibid.

10 Billy Parish, "Judge Withdraws Black Mesa Mining Permit." *Native American Times.* Jan. 8, 2010. http://nativetimes.com/index.php?option=com_content&view=article&id=2842:judge-withdraws-black-mesa-mining-permit&catid=56&Itemid=32.

11 Great Lakes Commission, "Great Lakes Basin Compact." http://www.glc.org/about/glbc.html.

12 Newmont Mining Corporation, "For Mining Firms, It All Comes Down to H2O." Adapted from *Waste & Recycling News.* http://www.newmont.com/features/our-environment-features/For-Mining-Firms-It-All-Comes-Down-to-H2O%20.

13 Newmont Mining Corporation, "Shaping Sustainable Development in Peru." Beyond the Mine: the Journey Towards Sustainability. http://www.beyondthemine.com/2009/?l=2&pid=240&parent=253&id=428.

14 See "California's Water-Energy Relationship," California Energy Commission, November 2005. http://www.energy.ca.gov/2005publications/CEC-700-2005-011/CEC-700-2005-011-SF.PDF.

15 Cost estimates for this project vary, but see for example Elaine Kurtenbach, "As US Debates Projects, China Builds Them," *Courier Post Online* (AP). Dec. 26, 2010, http://www.courierpostonline.com/article/20101226/BUSINESS/12260323/As-U-S-debates-projects-China-builds-them.

16 Michael E. Webber, "Energy Versus Water: Solving Both Crises Together." *Scientific American.* Oct. 22, 2008. http://www.scientificamerican.com/article.cfm?id=the-future-of-fuel.

CHAPTER 6

FUTURE SOURCES
OF WATER

In the previous three chapters, we've discussed how water may be used in the home, to grow food, and to make things in the future, but we haven't really talked about where that water will come from. In this chapter, we'll talk about where water comes from. What will the sources of water be in the future?

Just to be clear, we will be talking about new sources of water to be used by humans here on earth. Obviously, from a planetary perspective, there simply aren't going to be any *new* sources of water—but there may well be ways of taking currently unused or unsuitable forms of water and turning them into water that we can drink or use in other ways. One quick caveat: many people in the water industry include conservation as a new *source* of water—and certainly it is one of our best and cheapest ways to make more water available. How to use less water per person has been a central theme of the previous three chapters, however, so we won't discuss that here. Instead, this chapter is about the *actual* sources of new water that can be tapped.

FRESHWATER FROM THE OCEAN

There's no bigger topic when discussing new future sources of freshwater than desalination—making freshwater out of seawater. So we'll start there. An unimaginably large amount of water lies in the oceans. Our whole focus in this book is about the 2 or 3 percent of water on the planet that is *not* salt water, but there is still that other 97 percent of our total water out there that just happens to be salty. Humankind has understood for thousands of years that harvesting seawater would be a great way to find a plentiful supply of drinking water; we've just had trouble doing it.

While we're talking about the future here, it may be a good idea to start by taking a step back; a big step back, to the time of Moses. The book of Exodus, chapter 15, has a verse about the followers of Moses complaining because the water of Marah was too bitter to drink. So, Moses "cried unto the Lord; and the Lord shewed him a tree, which when he had cast into the waters, the waters were made sweet." Nice trick. Because we don't have Moses around, however, we're stuck with the regular ways of trying to get the salt out of seawater if we want to use it for drinking or to grow food.

The reality of removing salt from seawater is that technology today is essentially the same as it was some four thousand years ago, when the Greeks were already treating water by evaporating and filtering it, though they certainly had less success back then with the filtering. The same two techniques, however, are generally what's used today. The filters have gotten good enough to remove salt and are now referred to as reverse-osmosis membranes. Similarly, the boiling and distillation of water has gotten more efficient. We touch on some emerging technologies later, but the likelihood is that we'll be using those same two techniques in one form or another for many decades to come if we want to use the water from the sea. That's the bad news. The good news for fans of desalination is that gains in efficiency will make it more and more widespread in the coming decades. Efficiency is key for several reasons, but none bigger than the central quandary of the relationship of water and energy that we keep coming back to. The reason we keep returning to that question is that it is so central to the future of human activity on earth. If clean water were limitless, power would be much less of an issue because so much water is used now to aid in the generation of power. If electrical power were limitless, we could transport and treat water with ease. The availability of both energy and water, however, is quite limited, and the desalination challenge brings that fact into crystal-clear focus.

Perhaps the best way to illustrate that point is to look at the countries in the Middle East, the undisputed leaders in desalination around the world. These countries lie in a totally arid region, but one that boasts great quantities of fossil fuels. That's why nine of the ten largest producing desalination plants in the world are in that region: very little freshwater but relatively plentiful energy. While environmentalists aren't happy about it because of the power-plant emissions and by-product brine issues, the Middle Eastern governments have no problem using massive amounts of their energy to convert seawater into freshwater that can be used to grow crops to feed people. In addition to the energy consumption issues, the other big concern about either method of desalination comes from those who are concerned about its eventual environmental impact on the sea. To understand this, and how it may affect desalination trends in the future, we should take a quick look at how the process works.

In the case of thermal processes, the scientists determine the temperature needed to evaporate most of the water, and then they design a machine to heat the water just that much. In the thermal treatment of salt water—contrary to what many people believe—the water doesn't just all boil away to leave behind a pile of salt. Instead, a thermal desalination plant produces two streams of water: one with most of the salt removed that is suitable for drinking, and another stream with a much higher salt concentration than it had before.

In the case of filtering, the other technique for removing salt, seawater is pumped at high pressures through special membranes that collect the salt. This is called *reverse osmosis* (RO) because the membranes do the reverse of what generally happens in nature with salt and water, which is a blending, or osmosis. These membranes have very tiny pores in them through which the salt cannot pass. The tiny holes are so fine that they tend to hold back the water, too. In essence, the only way to get the water to flow through these membranes is to increase the pressure by using high-pressure pumps, thus adding to the overall energy consumption of the process. In addition, as you might imagine, those membranes quickly get clogged up with the salt, and so they need to be regularly cleaned. Just like with the thermal method, the RO plant produces two streams of water: one with less salt and one with more salt. As the price of manufacturing reverse-osmosis membranes has come down sharply over the past couple of decades, this method of desalination is gaining in popularity. In addition, RO treatment plants use less energy than plants that use the thermal methods, although they still use plenty. See Figure 6-1 (opposite) for a schematic summary of the recently completed Tampa Bay, Fla., desalination plant—a typical RO facility.

In both cases, the stream of extra-salty seawater, or brine, is typically returned to the sea near where the water was originally extracted. What else is there to do with it? This is what has many environmentalists concerned. Sea life has obviously evolved to thrive in salt water, but too much salt can be just as deadly for certain types of ocean organisms as seawater is for land animals and crops. Seawater can kill a human if ingested over time because the body does not have enough clean water to wash the salt away. The same is true for sea life if the salt concentrations are too high. Hence, environmental groups are increasingly arguing for more government control and regulation of brine stream disposal from desalination plants. Just as those environmental groups have been successful in recent decades in winning new protection for the land and air, they will likely have success in future decades in gaining new protections for the sea and sea life. But that will also mean that the already high cost of desalination will probably climb even higher.

As mentioned, the other problem from an environmental standpoint is that the discharged water is often quite hot—considerably hotter than the ambient temperatures of the ocean that it is being put back in to. Even water discharged from a reverse-osmosis treatment plant is hot enough to harm many forms of sea life near the discharge point.

So, what will desalination plants look like in the future? One good indicator is a plant that came online in 2009 in Barcelona. That plant, now the largest reverse-osmosis facility in Europe, takes care of the extra-salty discharge with what is now a unique solution

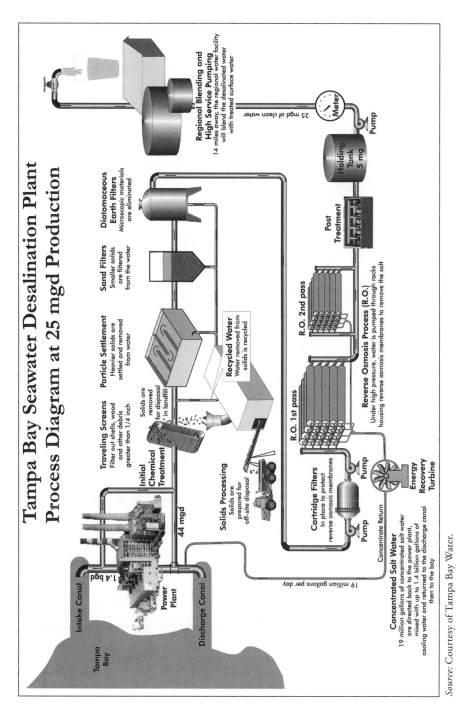

Tampa Bay Seawater Desalination Plant
Process Diagram at 25 mgd Production

Tampa Bay

Intake Canal

Power Plant

Discharge Canal

1.4 bgd

Traveling Screens
Filter out shells, wood and other debris greater than 1/4 inch

Initial Chemical Treatment

44 mgd

Particle Settlement
Heavier solids are settled and removed from water

Solids are removed for disposal in landfill

Sand Filters
Smaller solids are filtered from the water

Diatomaceous Earth Filters
Microscopic materials are eliminated

Recycled Water
Water removed from solids is recycled

Solids Processing
Solids are prepared for off-site disposal

Regional Blending and High Service Pumping
14 miles away, the regional water facility will blend the desalinated water with treated surface water

25 mgd of clean water

Meter

Pump

Holding Tank 5 mg

Post Treatment

R.O. 2nd pass

R.O. 1st pass

Reverse Osmosis Process (R.O.)
Under high pressure, water is pumped through rocks housing reverse osmosis membranes to remove the salt

Cartridge Filters
In place to protect reverse osmosis membranes

Pump

Pump

Energy Recovery Turbine

Concentrate Return

19 million gallons per day

Concentrated Salt Water
19 million gallons of concentrated salt water are directed back to the power plant, mixed with up to 1.4 billion gallons of cooling water and returned to the discharge canal then to the bay

Source: Courtesy of Tampa Bay Water.

FIGURE 6-1   SCHEMATIC OF A TYPICAL DESALINATION PLANT

that will likely become more commonplace. The Llobregat plant combines its salty discharge water with treated wastewater effluent from a nearby municipal wastewater treatment facility. While that treated wastewater is not suitable for drinking water, it is not salty. So, the warm-water discharge from the desalination plant is combined with the discharge of cooler water from the wastewater treatment plant. The combined discharge of the two plants is cooled during that mixing process, and then cooled further during a 3-mile trip through a pipe into the ocean, where it is discharged, with minimal impact, they hope, on the ocean life.

According to Barcelona municipal government, the plant is now producing the water for four and a half million people, nearly one-fourth of the Barcelona region's population.[1] Also, the municipal government will not be withdrawing water from the river Ter for the first time since 1967. Barcelona's water scarcity problems have in fact been so severe in the past that the city was occasionally forced to import drinking water via ships at a hugely prohibitive cost, one of the few cities in the world that has ever had to resort to actual shipping of freshwater.

Barcelona's operation gains some efficiency by combining two plants: a desalination plant and a wastewater facility. Similar combinations will probably become much more prevalent in the future. Many new desalination plants, or those in the planning stages today, most notably in Texas and in the Middle East, combine power plants with desalination facilities. In chapter 5, we discussed the two kinds of electricity generating plants: closed and once-through. In a once-through system, massive quantities of water are drawn in and heated up as they cool the plant. In a thermal desalination facility, massive quantities of water are drawn in and heated to produce drinkable water. Hence, it's not hard to see why it makes sense to use the "waste" coolant seawater from a power plant as the source of already heated water that more easily can be processed into drinking water.

Other more advanced desalination methods may also come into broader use in the future. One Canadian company, for instance, claims that it has figured out an entirely new way of desalinating water, not using filters or heat but small amounts of electricity and the principles of chemistry.[2] The exact method from Saltworks Technologies of Vancouver is a bit complicated for this book, but it takes advantage of the ionic charge of the two main parts of a salt molecule to attract the salt out of the water. Interestingly, the substance used to attract the salt is even saltier water. Seawater is 3–4 percent salt, but the process here uses water that's 18 percent salt. That super-salty water is held in a tank with ion bridges that are made from the same ingredients used to make Styrofoam packaging. At the end of each ion bridge is the seawater. A small charge gets run through and the result, says the company, is drinkable water and blocks of salt: no liquid brine at all. The company claims that the best place to set up this kind of plant would be next to an existing desalination plant, where it could use the effluent brine to make up the super-salty water needed for the process.

Other interesting and emerging new desalination technologies almost all require massive infrastructure and large-scale energy inputs that typically can only be met by the burning of coal, natural gas, or oil. There are other possibilities for the future of desalination, however, some high tech and some decidedly low tech, and some as old as the earth. Let's look at the high-tech possibilities first.

INNOVATIONS IN DESALINATION

Earlier we discussed nanotechnology, an emerging science that builds functioning systems at the molecular level and that can, for example, create clothing that can be cleaned without water. Most nanoscience today is related to the development of new medicines, but one group of scientists at MIT has developed a nanodevice that can purify water of not just salt but also organic material, particulates, and other contaminants and that can fit in a small backpack.[3] The heart of the device is incredibly small: 1,600 of them are spaced out on an 8-inch disc. The advantage of this kind of device is that, unlike reverse osmosis, there's no actual filter, so the unit itself doesn't clog up. It may get used someday for municipal-scale facilities, but the researchers say that the small size is something they think could be useful soon for remote homes and villages or for water treatment in the wake of disasters when municipal water is unavailable. Gallon per gallon, it uses about the same amount of energy as a large reverse-osmosis plant, but because of its small size, the device could be powered by rechargeable batteries or a small solar collector and produce four gallons of water an hour.

The biggest problem in turning seawater into drinking water, as we've seen, is the energy needed to remove the salt. What if that energy came from the sea itself? In the future, it just might. The US Army Corps of Engineers recently granted approval for a private company, Independent Natural Resources, Inc., to build a test facility a mile off the coast of Freeport, Texas. Here's how it will work: The platform at sea will use blocks filled with air inside of a large cylinder. As the blocks move up and down with waves and the tide, water will be drawn in, and the power of the waves would generate the electricity needed on the platform to drive a reverse-osmosis filtering system for the seawater. Because the intake is from the natural movement of the sea, the engineers say that it will not harm ocean life in the way that pump intakes can. The salty water effluent is not heated and will be discharged remotely and mixed with the tides, so the impact there will be minimal. The company's test begun in 2010 hopes to produce 3,000 gallons per day of drinking water, but a spokesman for the company said the design for one platform could easily produce 60,000 gallons of water per day for municipal or agricultural uses, all of it without any external power inputs.[4]

These innovations, and many more examples similar to them, are the kinds we expect to see much more of in the future. But perhaps the biggest desalination trend of the future will be the simple utilization of the sun to provide the energy required.

Collecting the condensed water from the evaporation of seawater heated by the sun is something of a holy grail for desalination supporters: creating freshwater without a carbon footprint. Dozens of academics, inventors, and activists have developed various forms of devices to address this challenge, with varying degrees of success. The essential problem is that such a system would work best in a desert that has an easy supply of seawater and with a population that has an economic interest in a successful result. But the problem is that the biggest and driest deserts in the world are generally far from the ocean; furthermore, few people in places like the middle of the Sahara Desert have the economic resources to build innovative plants like this. Hence, the best hope for taking advantage of the power of the sun to remove the salt from the seawater lies in the economic needs of those far away, sometimes on another continent. One consortium in Europe is pushing right now for a system that would pipe seawater into a low spot of the Sahara and evaporate the water to create an oasis in the desert and a new source of electricity that would be exported to Europe.[5]

While some distributed or decentralized lower-tech methods may catch on, one of the bigger trends that will likely grow in coming decades is *concentrated solar power*. The most common advanced solar power technology today is the conversion of the sun's rays directly into electricity using photovoltaic cells. Less common is concentrated solar power, which is a system of focusing the sun on a single point using an array of mirrors to create very high temperatures. Through that single point, water is pumped, and the resulting superheated water can then be used to generate electricity. In addition, the distillate can be condensed into a clean drinking water. A version of such a system is in use right now in the Mojave Desert of California,[6] and while it's a hundred miles from the ocean, it is now serving the power needs of those who live in a town northeast of Los Angeles.

The key to growth in this kind of solar energy and desalination may be, counterintuitively, software. It seems certain that the price of water will climb, and the only thing that's generally predicted to decrease in the future is the cost of computing power. Some new companies are trying to take advantage of that by designing large and complex solar arrays that train hundreds of mirrors in the exact direction of the sun so that the full intensity of the sun's warmth can boil water quickly in a tank or tower where the heat is focused. Current arrays focus the sun's heat but lack that kind of pinpoint precision. Companies like eSolar, Inc., are hoping to do the difficult job of making advanced geometric calculations for thousands of mirrors, and doing it every couple of seconds, to track precisely the sun and be able to boil cheaply the water to drive steam-powered turbines. These sorts of concentrated solar power projects could produce steam for power—or desalinated water—at very attractive operating costs, but they obviously entail huge up-front capital spending first.

Interestingly, the technology at the actual turbine hasn't changed much; eSolar has the first concentrated solar power generator online in the United States in the town of Lancaster, Calif., near Edwards Air Force Base, and the steam produced by the new tower goes into a refurbished 1947 power-generating turbine. Right now, this technology is used primarily for power generation, but the adaptation to using a system like this for desalination will be coming soon enough.

The desalination of seawater will continue to be one means, albeit expensive and generally last-resort, of stretching our freshwater supplies. As we've seen, it's not a simple or slam-dunk alternative: it's expensive, it comes with various environmental issues or questions, and it's generally only going to be considered in heavily populated coastal areas that have completely run out of alternative water sources and that have access to relatively cheap energy. That said, however, desalination *is* one of the most active areas of technological research and development, and costs will undoubtedly continue to come down. And it's not strictly a coastal solution for parched energy-rich Middle Eastern cities; some inland areas of the United States are also looking at desalinating briny groundwaters. Phoenix is looking at alternatives here, and El Paso has already built one of the world's largest inland desal plants as a supplementary source of drinking water. Desalination will be an important but small overall contributor to the world's supply of freshwater.

OTHER NEW SOURCES OF WATER

So, it's pretty clear that the ocean is by far our biggest untapped source of new freshwater for drinking and feeding to crops. But what are some of the others? Can we really consider existing sources—rivers, lakes, underground aquifers, and so on—to be *new* sources? Perhaps; we'll have more on this later. Let's first consider another resource we take for granted: the air we breathe. It may turn out that our air can be a new source of water, too.

Fans of *Star Wars* may remember that in the original movie, Luke Skywalker's uncle wanted a droid that could work with "moisture vaporators." They were living on a desert planet, and George Lucas conceived of a device that would do the opposite of a humidifier that puts moisture into the air and instead would suck water out of the air. Of course, that's not really science fiction; dehumidifier devices in fact have a long history. Archaeologists believe that structures dating back to the Byzantine era were condensers, or air wells, that promoted the condensation of moisture from the atmosphere, with stones that were cooler than the dew point for water. Similar types of collectors have been used in various situations throughout human history. Right now, for example, hundreds of people live on farms irrigated with water harvested from the air around Lima, Peru, which gets very little rain but is covered in dense fog for about nine months of the year.

Solutions to collect water from the air have been around a while, and there's even an International Organization for Dew Utilization.[7] A number of dehumidifier technologies, or atmospheric water generators as they are now called, have recently been promoted to water industry investors. Still other innovations are in the works, including giant collectors that would be built near the seashore to capture the water from ocean breezes.

Other proposals exist for moisture collectors that can work even in the driest climates. The *New York Times* reported on a beetle that survives on the moisture in the air above a desert, where there isn't a lot of water in the air.[8] The Stenocara beetle has a unique series of peaks on its back that collect water and troughs that funnel the water to its mouth; scientists believe they may be able to fashion a plastic sheeting with similar characteristics to collect water. These scientists estimate that they can extract as many as 12 gallons of freshwater per day from the air in most regions by operating a unit that measures about 300 square feet and can be installed on top of a building.

And of course, by now most of us have heard of various and seemingly more far-fetched plans: to capture freshwater from the melting glaciers of Greenland, or somehow harvest the polar icecaps and drag them to Barcelona, Mumbai, or other tropical cities with water shortages. Recently, the town of Sitka, Alaska, garnered international attention when it announced plans to consider the export to India of millions of gallons of glacial meltwater held in a reservoir near the coast. More than a decade ago, a number of companies were lining up to transport freshwater from British Columbian rivers in massive tankers to various points around the world, until the government there decided to ban the potential practice. Aquarius Water Trading and Transportation Limited of Greece already operates a fleet of massive polyurethane bags—holding as much as a half million gallons—that it tows by tugboat to various Greek islands suffering from droughts in the summertime.[9] As mentioned earlier, Barcelona has already had to resort to importing water in converted vegetable oil tankers to help solve short-term water crises. Bulk international transport of freshwater may seem a little far-fetched today, but as prices continue to rise, we may see countries like Canada, Norway, or Iceland begin to examine the economic possibilities in the future.

The problem with most of these innovative technologies is scale. While they may work for handfuls of people in geographically remote regions, they probably won't be able to help make a serious dent in the water needs of a highly urbanized planet with more than six billion people. The unit that collects water from an area of about 300 square feet may work great in the countryside but obviously won't in the cities where we will see most of the population growth around the world in the coming decades. Twelve gallons a day won't supply even the barest needs of a single person, and in a city like Mumbai, with perhaps twenty-five million souls, there obviously isn't even remotely enough surface area to allow these kinds of technologies any role at all in clean drinking water provision. Every little bit helps, but the air pollution in big cities would also foul drinking water gathered from the air.

GROUNDWATER MANAGEMENT

So, that leaves us with the water that either comes from underground or gets collected from our rivers and lakes, already the current sources of most of our water. Let's take a closer look at those sources, starting with water from underground aquifers.

This is a book about the future, so a discussion about the extraction of water from underground aquifers will be relatively short—simply because the future just isn't very bright for the unsustainable pumping of underground water. The situation varies from region to region, based on human history, geography, and geology, but in many areas there's not much future at all. In general, most of the best places around the world to pump and utilize underground water have long ago been tapped and in many cases have already been egregiously overused. In some cases, aquifer waters are truly a renewable resource, being recharged by rain and snowfall that gradually infiltrates and seeps underground to a hidden basin, where it is stored. From there it can be drawn up and used by people on the surface. These types of recharging aquifers will undoubtedly continue to be exploited into the coming decades. Unfortunately, however, this kind of aquifer situation is not very common. In most areas, we are utilizing water that has been underground for thousands or even millions of years, and we are using it at rates far in excess of the rate at which the water is being replenished. Because in this type of situation we are essentially harvesting a nonrenewable resource, this practice is often referred to as the *mining* of groundwater.

One of the most studied groundwater basins in the world is the Ogallala Aquifer, the enormous body of water that underlies the Great Plains, spanning parts of eight states, including Nebraska, Kansas, Oklahoma, and Texas. With advent of rural electrification and the development of relatively inexpensive electric pumps in the middle part of the last century, irrigated agriculture became possible in areas that were formerly too arid to really support much in the way of agricultural crops. Over the following several decades, water was drawn out of the ground at an increasingly furious clip. And even though most of that withdrawn water is used for agricultural irrigation, a lot of the water that doesn't go into nourishing the crops doesn't return to the aquifer: the land is so dry that most of the water not needed for the plant either evaporates or runs off into surface waterways. As a result, this once vast aquifer has been seriously drawn down and depleted and in places is close to being exhausted. Irrigation wells have had to be drilled deeper and deeper, and pumping costs have risen proportionately.

Today, across western Nebraska and Kansas and in Colorado, we see wider and wider swaths of land that have been agriculturally productive for several decades being returned to the natural semiarid state that they were in before widespread groundwater pumping occurred in the mid-1900s. The water is either gone, is prohibitively expensive to pump, or has been sold to burgeoning and thirsty cities. This is a trend that's likely to continue and accelerate.

In some cases, even if somehow a source of water were available on the surface to recharge an aquifer, it might be impossible to actually do so anyway. The US Geological Survey estimates that Las Vegas is now about 6 feet lower than it was before World War II, and the reason is that the ground itself has started to subside as more water is drawn out of the underlying aquifer.[10] As the water-bearing salty or sandstone layers lose their contained water to human usage on the surface, they may become structurally weaker, then may gradually collapse and compact. Crushed into a denser mass, the underground layers have permanently less room in which to store water, even if there were water available to store, which there typically isn't. The situation is even more serious in parts of California, where reports indicate that in many places surface levels may have sunk 30 feet or more.[11]

Similarly, Mexico City, one of the largest cities in the world, has experienced as much as 25 feet of subsidence over the past hundred years.[12] If this kind of subsidence were uniform, it might not be a huge problem on the surface, but the underlying geology is variable, and so is the subsidence after the underlying rock layers have been depleted of water. Boulevards have cracked, historic churches have tilted and fallen over, and underground gas and water utility pipes have been twisted and broken. In Mexico City's case, this is in addition to the vast poverty, earthquakes, and severe air pollution problems that the city already suffers from. This type of subsidence and compacting is permanent, and the ability to recharge or reutilize such original aquifers is lost forever.

While the Great Plains and the dry Southwest understandably get much of the attention, those regions certainly aren't alone in pondering a future with a much reduced potential for groundwater mining as a source for human consumption. In the United States, even the relatively humid and wetter regions of the Southeast are realizing that they are removing water from underground sources at a rate two or three times as fast as the rate at which those aquifers can be recharged. Unless consumption is drastically curtailed, wells will have to be dug deeper and deeper and more electricity will be needed to pump the water up greater distances, and ultimately those aquifer resources will be effectively exhausted. Many major urban areas are already desperately close to this type of reckoning day.

Coastal regions also face another challenge in terms of groundwater management: as fresh groundwater is unsustainably pumped out of the underlying rock, seawater may seep into those aquifers adjacent to the coastline. Not only may the groundwater simply be overpumped, but as more water is pulled out, seawater can migrate into the newly drained underground aquifer areas. That leads to coastal users beginning to taste salt in their well water.

A simple solution exists to the problem of seawater intrusion: reducing the amount of water pumped up from underground to begin with. While technically that sounds easy, practically it's not easy if the aquifer is the prime source of water for a rapidly growing city. The USGS reported the results of an ongoing project begun in 1978 in the New Jersey Coastal Plain, an area of three counties in New Jersey on the coast about halfway between New York City and Atlantic City.[13] Once the state started mandating reduced withdrawals from the aquifer, the water levels returned to previous standards within fifteen years. (While that solution sounds easy, it did not come with an actual reduction in water usage: water sources were switched from groundwater to newly constructed reservoirs fed by surface water.)

Seawater intrusion is an issue all up and down both of our main coasts and also along the Gulf of Mexico. For example, the town of Brunswick, Ga., has been pumping groundwater for its residents since after World War II. In the 1950s, local officials discovered that their pumps were starting to bring up seawater. By 1962, the wells yielding salty water covered an area less than half a square mile and none of them had high concentrations. By 1991, both the concentration of salt and the area that was contaminated had increased—wells in a 3-square-mile area now pumped water with salt concentrations that rivaled seawater. The concentrations and the area affected continue to grow, but unfortunately the behavior of the human population on the surface hasn't really changed much. A study done in 2006 by Georgia Southern University found that while a majority of residents agree that saltwater intrusion is a threat, very few residents were doing anything to conserve water.[14] The biggest water users—the pulp and paper companies that make products from the nearby forests—have been contributing to studies to figure out the future of their ability to extract water from underground. One of the companies, Rayonier, Inc., is the largest user of pumped water in the state, according to the Georgia Water Coalition. In an interview, a spokesman for Rayonier said, "There's no shortage of water; there's just a shortage of cheap water."[15]

The second problem related to seawater intrusion may stem from the effects of global climate change. If sea levels rise even a small amount, the areas where seawater impinges on freshwater aquifers could dramatically increase and spell the end of groundwater extraction in some areas close to oceans. Rising sea levels could also increase salinity levels farther upstream in the estuaries of those rivers flowing into the oceans—from which many coastal cities draw their municipal water supplies. Recently, a group of state agencies from California, Oregon, and Washington, along with the National Oceanic and Atmospheric Administration, the US Geological Survey, and other federal agencies announced that they would evaluate all the science to date, and try to project sea levels over the coming decades.[16] That study is due out in 2011 or 2012.

Around the world, the story is pretty much the same: pumping too much water from an aquifer means that either it is gradually depleted, that it gets contaminated with seawater (if it is close to the coast), or that it subsides more quickly than it can be replenished, and as a result recharge is difficult or impossible. In Saudi Arabia, where the government desalinates more seawater per capita than any other country on earth, they still rely for more than two-thirds of their water on nonrenewable groundwater sources. Since these aquifers are drying up, the country has abandoned its program begun in the 1970s of trying to grow wheat in the desert. The program succeeded in growing enough wheat to feed the people, and it even exported some wheat in years of crisis to other countries. Now, however, as groundwater becomes scarcer and dearer, that era is over, and the country has inked deals with countries such as Ethiopia and Sudan to provide wheat in the future. The government plans to eliminate all wheat production by 2016.

In China, it's a similar story: the groundwater under the vast North China Plain is being removed so rapidly that many wells have already essentially gone dry. As a result, wheat production in that area has dropped by at least half. In India, the water table is dropping so fast that many of the small farmers are starting to encounter dry wells, and they don't have the money to dig deeper. Larger industrial users may have the money, however, and as they sink wells ever deeper, many farmers, with nothing else, end up using industrial wastewater to irrigate crops. The list of challenges in groundwater management, overuse, and exhaustion go on. As we said, there's just no future in mining underground water.

OTHER APPROACHES

After the oceans, the air, and underground, there just aren't many new sources of water. Most of the water on the surface is already used one way or another. And in many parts of the world, for example the eastern industrial areas of China, that surface water is so contaminated that it's virtually unusable. There's just not much water left on the surface. We'll write more about storage of water in the next chapter, but it's clear that the era of big reservoirs and dams in the United States is probably coming to a close. Do other possible ways exist to use our surface water resources differently?

Maybe so. A few big-thinking entrepreneurs have proposed one plan or another over the years to move water out of the Mississippi and into the West. It's understandable why people dream that dream: the average flow of the Mississippi River is forty times that of the Colorado River. While every drop of the Colorado gets used and water scarcity is one of the most significant economic issues in the Southwest, the federal government has spent billions on dams, levees, and other measures to keep the Mississippi from flooding, and still the Big River does pour over its banks regularly.

So, why not just capture and move some of that excess water? The excess Mississippi waters are only causing problems right now, and maybe they could better be used somewhere else. Pat Mulroy is no wild-eyed entrepreneur; she's the longtime chief of the Southern Nevada Water Authority, the water utility that serves the desert oasis of Las Vegas, and is often mentioned as one of the most powerful people in the world of water. She gave a speech to the Brookings Institution in 2009 saying that the federal government should study the idea of tapping the Mississippi and using its floodwaters to recharge the Ogallala Aquifer under the Great Plains.[17] It might take a while, she said, but once built could create a sort of domino effect. Right now, many farms, factories, and people along the Front Range of the Rocky Mountains in Colorado use water diverted from the Colorado River on the Western Slope of the mountains. If those transmountain diversions could be eased, then that much more water would be left in the Colorado River for Las Vegas and other downstream cities.

So far this is only an idea, and no major studies have been announced to more carefully evaluate the plan. A project of this scale would obviously be a hugely contentious and politically explosive issue: it would require the rewriting of interstate compact laws and practices that have existed for a century, so it's no easy thing. But as the situation gets tighter, we'll naturally be forced to look at more radical solutions.

A similar proposal in drought-plagued Australia was recently abandoned. A group of water companies wanted to build a pipeline from the island state of Tasmania more than 200 miles under the Bass Strait to Melbourne, Victoria, on mainland Australia. As with the Mississippi, Tasmania's rivers have plenty of excess water that flows into the sea every year, and the engineering challenges wouldn't be as severe as the Mississippi-to-Ogallala plan. The political desire to build such a project, however, appears to have waned now, as the cost was prohibitive—and the Australian government seems more focused on other supply strategies, including desalination. There's also the small matter that the Tasmanian government is against the plan, saying that while they do have plenty of water on the western side of the island, the eastern side is dry. Officials on that island say they want every one of the farmers on the eastern half of the island to have all the water they need before diverting any water across the strait to Melbourne.[18]

Just as with the Mississippi plan, however, a massive diversion from one watershed to another might be possible if the costs of water continue to rise. So whether it's one of those two plans, or one of dozens of others around the world, the plans today that are currently the province of big thinkers will become the plans that only make sense when the costs of water exceed the costs of the status quo.

Right now in the Western United States, numerous interbasin and interstate pipeline projects are on the books, all geared to moving around massive amounts of water from areas of more naturally abundant water to the population centers. But given the economic costs, the political sensitivities, and the potential environmental side effects, it is unlikely that many of them will ever actually be built. As we'll discuss later, as water gets more and more expensive, it's not out of the realm of possibility that somewhere in the future, we'll see people migrating back toward water, rather than people figuring out how to move all the water toward them.

℘

[1] Mireia Díaz, "The Llobregat Desalination Plant Will Ensure 24% of the Metropolitan Region's Water Consumption." Council of Barcelona News. July 20, 2009. http://w3.bcn.es/V01/Serveis/Noticies/V01NoticiesLlistatNoticies Ctl/0,2138,1653_35144087_3_924336390,00.html?accio=detall&home=Ho meBCN&nomtipusMCM=Noticia.

[2] *Economist*, "Cheaper Desalination: Current Thinking." Oct. 29, 2009. http://www.economist.com/node/14743791.

[3] David Chandler, "A System That's Worth Its Salt." *Massachusetts Institute of Technology News.* March 23, 2010. http://web.mit.edu/newsoffice/2010/desalination-0323.html.

[4] Martin LaMonica, "Wave-Powered Desalination Pump Permitted in Gulf." Green Tech. *CNET News.* May 28, 2010. http://news.cnet.com/8301-11128_3-20006269-54.html.

5 Alok Jha, "Seawater Greenhouses to Bring Life to the Desert." *The Guardian*. Sept. 2, 2008. http://www.guardian.co.uk/environment/2008/sep/02/altern ativeenergy.solarpower#.

6 Renewable Energy Development, "Solar Energy: Mojave Solar Park (CSP)." March 19, 2008. http://renewableenergydev.com/red/solar-energy-mojave-solar-park-csp/. See also http://www.esolar.com/.

7 International Organization for Dew Utilization, http://www.opur.fr/index. htm.

8 Kenneth Chang, "Like Water Off a Beetle's Back." *New York Times*, June 27, 2006. http://www.nytimes.com/2006/06/27/science/27find.html.

9 Jeffrey Rothfeder, *Every Drop for Sale: Our Desperate Battle over Water in a World About to Run Out*. Penguin, 2004.

10 S.A. Leake, "Land Subsidence from Ground-Water Pumping." Workshop on Impact of Climate Change and Land Use in the Southwestern United States, July 1997. US Department of the Interior, US Geological Survey. http:// geochange.er.usgs.gov/sw/changes/anthropogenic/subside/.

11 USGS Groundwater Information, "Land Subsidence in the U.S." USGS Fact Sheet 165-00. 2009. http://water.usgs.gov/ogw/pubs/fs00165/.

12 See for example Sam Dillon, "Mexico City Journal: Capital's Downfall Caused by Drinking . . . of Water," *New York Times*, January 29, 1998. http://www.nytimes.com/1998/01/29/world/mexico-city-journal-capital-s-downfall-caused-by-drinking-of-water.html.

13 Vincent dePaul, Robert Rosman, and Pierre J. Lacombe, "Water-Level Conditions in Selected Confined Aquifers of the New Jersey and Delaware Coastal Plain," 2003. New Jersey Department of Environmental Protection. USGS Scientific Investigations Report 2008–5145. http://pubs.usgs.gov/ sir/2008/5145/.

14 Godfrey A. Gibbison and James Randall, "The Salt Water Intrusion Problem and Water Conservation Practices in Southeast Georgia, USA." *Water and Environment Journal* 20, no. 4 (12, 2006): 271–281.

15 *Atlanta Journal-Constitution*, "Water Spat Persists: Debate Continues on Saltwater Threat." August 7, 2005.

16 California Energy Commission, "California State Agencies Collabora-tive Research Project National Research Council (NRC) Sea Level Rise Study." http://www.energy.ca.gov/2010publications/CAT-1000-2010-005/ Research_Collaboration_Case_Studies/Sea_Level_Rise_Study.pdf .

17 Henry Brean, "Mulroy Advice for Obama: Tap Mississippi Floodwaters." *Las Vegas Review Journal*, Jan. 12, 2009, http://www.lvrj.com/news/37431714.html.

18 Felicity Ogilvie, "Tasmanian Government Rejects Pipeline Plan." PM (Austra-lian Broadcasting Corporation). Feb. 4, 2009. http://www.abc.net.au/pm/ content/2008/s2482512.htm.

CHAPTER 7

THE FUTURE OF
WATER STORAGE

It's hard to overstate the significance that dams have had in the economic development of the United States over the past hundred years. During the middle decades of the twentieth century, we built about one dam every day in this country. And elsewhere around the world, many significant dams have been and continue to be built with financing from the World Bank, which is largely financed and controlled by US interests. But nothing matched the United States for the complete saturation of dams built in the mid-1900s.

The current century will be much different. In the United States today, most of the news about dams is how they are being decommissioned and torn down. In the past, dams were not just the symbol of progress; they in fact powered much of that progress. Now progress is increasingly being defined as returning rivers to their native, original, and free-flowing ways, restoring balance to the environment.

So far in this century, the construction of dams is a story owned by other parts of the world, especially Asia and Africa. These dams are no longer going up with World Bank financing, however. The future of dams around the world is increasingly being written by the Chinese, who are either building or financing nearly all of the significant dams in these parts of the world.

Why are dams so important? Well, for several reasons. First, they help us control floods. Second, they can provide cheap power once they are built. And third, people need water all year round. In most regions, water flows seasonally, so we will always need some form of water storage; the reservoirs behind dams provide this storage. An increasing trend in the future both in the United States and around the world may be underground storage. We'll get into that later, but let's begin with the biggest story in dam building right now: China.

CHINA'S VAST DIVERSION PROJECTS

While the dam-building era is largely over in the United States, the public perception of the history of dams is in general a positive one. Probably the most iconic of our dams is the Hoover Dam, credited not only with helping the country work its way out of the Great Depression, but also with generating the electricity and the industrial base that helped to power the United States to victory in World War II.

In China, the stories aren't so popular and seem to be purposely hidden from view by government officials. It's easy to see why. Take, for example, the Yellow River. It's called the Yellow because it often runs yellow owing to the high volume of sediment it carries, now more yellow than ever because it drains a lot of deforested land. In 1931, the river flooded and the death toll was estimated to be somewhere between one and four million people. The government didn't want that to happen again, so it built a series of levees and dams along the river. In 1938, Chiang Kai-Shek ordered the levees and dams bombed in an effort to stop the invading Japanese. That single act of sabotage probably killed more people than any other in world history—reportedly nine hundred thousand but likely many more—virtually all of them Chinese citizens who were purportedly being protected.

In just the past couple of years, the largest hydroelectric dam in the world has been completed, and the reservoir behind the Three Gorges Dam is filling with Yangtze River water, forcing the relocation of between two and six million people from thirteen cities,

FIGURE 7-1   THREE GORGES DAM ON THE YANGTZE RIVER, CHINA

one hundred forty towns, and three hundred twenty-six villages. Most of the residents downstream of the Three Gorges Dam have never heard the story of what happened in 1938 on the Yellow River, because the government censors that story. And it's easy to see why it might make some people downstream jumpy.

Because of the state-controlled media, those people have also probably not heard that even the US military is apparently keeping an eye on the enemies of China. Some other country might not have the strength to take on the Chinese army but perhaps could launch a terrorist attack on just that one dam. If, for example, the Taiwanese destroyed the dam, it could wipe out as much as 10 percent of the nation's power supply and endanger perhaps three hundred million people who live downstream of the new reservoir. That is, the population in the floodplain of that dam is about the same number of people as the total population of the United States.

The quantity of water in the reservoir behind the dam is so massive that scientists think that its sheer weight may alter the way earth spins on its orbit. (Interestingly, similar calculations were able to be made with the construction of the Hoover Dam and all of the cement it took to create that edifice.) A number of seismic events have already occurred, spawned by the reservoir. These events have triggered landslides nearby and caused cracks in the dam that have required extensive repairs, but still officials insist that the dam is sound and the population need not worry.

And there are other concerns. The dam has been advertised primarily as a way to generate power. It may create as much electricity as fifteen nuclear power plants, but the actual numbers are not clear, and the dam is quite remote from the key markets. The project has also been billed as a way to control floods, although many experts blame recent flooding more on large-scale deforestation. Finally, the dam planners hope to harness water for irrigation, but a number of issues make it clear that it will not be that great a water source; for one thing, some forty cities and more than four hundred factories send untreated wastewater into the new reservoir. A skeptic could well say that a massive septic tank is now flooding what had been some of the most picturesque area of the country, along with thousands of archeological sites, some of them six thousand years old. One recent report indicated that the garbage and debris collecting on the surface of the reservoir was so thick and compacted that people could actually walk on it.

The dam has cost $30 billion, or perhaps as much as $75 billion; exact numbers are frustratingly hard to pin down. But if Chinese officials have their way, the Three Gorges Dam is only the beginning. China has also announced the great South-to-North Water Diversion Project, which will consist of three major pipelines moving water over hundreds of rivers and dozens of rail lines. The project will move water past hundreds of cities, pulling water out of the Yangtze and its tributaries to deliver it to the relatively parched northern parts of the country. Construction has already begun, and even the official cost estimates put this project at twice as expensive as the Three Gorges Dam. The original plans for the project did not call for pulling water from the reservoir, but as water needs increase, the latest indications are that water from Three Gorges may be included in the massive diversion plan.

The bigger question mark regarding the whole Three Gorges Dam and these related projects may be, at least from the water perspective, whether they are really needed at all. Water waste and horrific water pollution problems are both huge issues in China. Agricultural and industrial users have yet to invest in or adopt any of the standard water conservation strategies and many of the water treatment processes being implemented elsewhere—and hence, a high percentage of the water is simply wasted. The same trends are generally observed in the big cities, where the Chinese population is rapidly aggregating. But even though the government closely regulates the use of gas furnaces and air conditioners, the state essentially gives water away for free without restriction. Even Beijing, which had a massive infrastructure improvement program before the 2008 Olympics, still suffers from as much as 50 percent loss or theft of water from its municipal water distribution system.

And as if these two massive water projects weren't enough, China is just finishing construction on several dams in the southern part of the country that will block several rivers that eventually turn into the Mekong River, running through Thailand, Laos, Cambodia, and Vietnam.[1] The area below the dam boasts an extraordinarily lush and diverse habitat, and already those downstream say that the water flow and silt deposits are down to the lowest levels ever recorded. Because of these trends, the entire region could be in ecological peril. This dam is just one of a series of eight dams planned for the region, two of them among the largest dams in the world. All in all, as shown in Figure 7-2, China has far more dams than any other country in the world.

GLOBAL REACH AND CONFLICT

And there's more. China is also proposing a similar project in the Himalayas: building a dam so massive that it would produce 1.5 times the electricity that the Three Gorges Dam generates. However, the dam would also essentially cut off massive amounts of water flowing into India and Bangladesh from that crucial watershed. This project has not yet been actually confirmed by government officials as we write, but Tibetan activists say they've discovered plans for the dam and the power lines that would tie it into the national Chinese power grid. Indians and Bangladeshis are understandably concerned that this dam would represent a huge power grab for water.

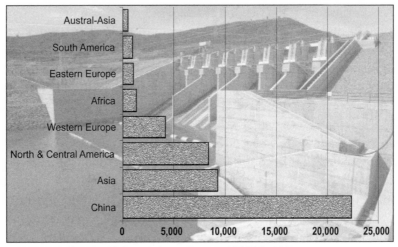

Source: Estimates from World Commission on Dams and ICOLD 1998.

FIGURE 7-2 DISTRIBUTION OF LARGE DAMS BY REGION

The Chinese have certainly shown a willingness to fight over water. Armed conflicts have taken place over water between internal Chinese provinces, the most recent between two local governments situated about halfway between Beijing and Shanghai that actually bombed water canals and other facilities, leading to widespread destruction of those facilities and nearly a hundred deaths. This is not a historical anecdote from ancient times: this conflict happened in 1999. In 2000, another six people died when officials in one province blew up a channel designed to divert water to another province. (We should also note that, closer to home, the states of Georgia and Tennessee are also currently battling over water from the Tennessee River, but they are fighting with politicians and legal briefs, not bombs.) So, if the Chinese are willing to bomb each other because of water, it's not beyond the imagination that they would go into war against another country in order to keep their water in the country. If the superpowers of India and China ever do go to war, chances are it will be over water.

Much more could be said about water storage inside China, where the president, Hu Jintao, is a trained water engineer whose first job out of college was working for Sinohydro, the Chinese engineering conglomerate.

Perhaps just as telling for the future as what's going on inside China is what the Chinese government is doing in other countries. In Ethiopia, for example, a planned dam on one of the last great stretches of free-flowing river in the world, the Omo, has been on hold for years because of concerns about the fragile ecosystem it supports, not to mention the half-million indigenous people downstream who depend on the regular flow of the river for survival. The Omo River valley is also a designated World Heritage Site and has provided fossil records of some of the earliest hominids. Archaeologists fear all that will be lost forever in the inundated area behind what's to be called the Gibe III dam. The World Bank has refused to fund the project, but the Industrial and Commercial Bank of China has no such concerns and recently offered a loan of approximately $500 million to get the dam built. A Chinese company will build the equipment for the dam. China will also be first in line to benefit from the agricultural output made possible by newly irrigated land and from the raw materials like copper and other metals that will now be economically extractable given the cheap electricity from the dam.

Similar stories could be told in Gabon, the Republic of Congo, Mozambique, Nigeria, Sudan, and Zambia: China is building dams in all those countries, along with other infrastructure improvements, while inking deals to own and operate a broad array of mines and farms. Given the relatively poor management and conservation techniques the Chinese use in their own country, the future looks bleak for the careful storage and use of freshwater in Africa.

As we've said, this is not a book about what we think *should* happen in the future. Instead, it's a book about what we think *will* happen. Unless a dramatic change happens in the government or economy of China, dam building in Asia and Africa will likely see an expansion over the next several decades similar to what we saw in America over the past hundred years.

THE DECLINE OF US DAMS

The picture in America and Europe couldn't be more different. The story here in the United States in the coming decades will be all about dams coming down, not going up. Given the age of many of the dams in the United States, this isn't really a surprise. The US Federal Emergency Management Agency estimates fifty-eight thousand large dams will essentially *expire* by 2020—exceed the life span planned for them when they went up in the twentieth century. This means that we have to either repair and rebuild the dams, or we have to tear them down. Given the costs and government approvals needed to rebuild a dam, many dam owners are choosing to just tear them down and return the rivers to their regular flows. But this, too, is far from easy.

One example of such a decommissioning is the Marmot Dam on the Sandy River in Oregon. The dam was a hundred years old in 2006, and the local utility that owned the dam decided to demolish it rather than try to restore it. Among other considerations, in order to meet the requirements of the Endangered Species Act the utility would have had to develop a plan to help ocean salmon get upstream past the dam. The costs of compliance were just too high, so the utility decided to remove it. Workers built a temporary coffer dam upstream to give themselves some room to work, and then with dynamite and backhoes removed the original 50-foot dam in 2008. When they were finished, they turned off the pumps that were diverting the river. Engineers had decided that rather than try to scoop out the sediment that had collected over the years behind the dam, they would just let the river naturally wash it all downstream.

What happened next surprised all the engineers and water experts, who had predicted that the remains of both the original and temporary dams, and all the sediment that had collected over the hundred-year life of the dam, would take perhaps two or more years to wash downstream. Instead, a relatively minor rain came and washed everything as much as 2 miles downstream in one day. The very next day, spotters saw salmon swimming upstream. The lesson learned will undoubtedly be repeated hundreds of times in the coming decades in the United States: rivers can very quickly return to their natural state once the man-made blockages are removed. That's good news.

A similar story was played out on the other side of the country, in Maine, where in 1997 the Federal Energy Regulatory Commission (FERC) refused to grant a renewal license to operate the Edwards Dam on the Kennebec River, ruling that the power produced by the dam fell short of justifying the environmental harm that it caused. The dam's operators had wanted to get a new license for the structure, which had been blocking the river in various forms since 1837. No government agency had ever refused a permit before, so the operators of the dam may have been surprised, to say the least, when FERC said no.

So in the summer of 1999, the Edwards Dam came down, and the river ran free for an additional 17 miles for the first time in one hundred and sixty years. As with the demolition in Oregon, the natural habitat quickly returned, with fish spawning and native grasses and other plants taking hold once again. A waterfall that had been submerged became visible again, and is now a sort of tourist destination. That was the first dam FERC had ever ordered removed, but it wasn't the last. The coming decades will certainly see dozens or maybe even hundreds more, whether the result of an order from a federal agency or just economic circumstances. A 50-foot, hundred-year-old dam simply costs more to maintain than it does to tear down.

But what about the big dams? Ever since publication of Edward Abbey's cult book *The Monkey Wrench Gang* in 1975, people have talked about removing the Glen Canyon Dam on the Colorado River. In the 1970s, it was the wild speculation of radical environmentalists. Now, talk about removing the dam comes from people such as Dan Beard, former commissioner of the US Bureau of Reclamation, who points out that removing the dam would be the cheapest and easiest solution to environmental and other problems in the region.[2] To be clear, the dam won't be coming down anytime soon; for one thing, a vibrant tourist economy now depends on the reservoir, called Lake Powell.

But as a water storage tool, the "lake" above Glen Canyon Dam—like all other dams and reservoirs—is not all that efficient or as long-lived as a water storage tank. First of all, it loses as much as 10 percent of the water on the surface to evaporation each year in this desert region. And second, sediment is building up on the floor of the reservoir, sediment that used to be transported downstream by the more rapidly flowing river. It is estimated that the equivalent of thirty thousand dump-truck loads of sediment are deposited into Lake Powell by the Colorado River every day. The intended storage capacity of the reservoir is gradually shrinking: on the top by evaporation or drought, and on the bottom by sediment buildup. Figure 7-3 shows a recent estimate of the percentage of reservoir storage capacity actually being used for the two largest reservoirs in the Southwest, Lake Mead and Lake Powell. Some observers have sarcastically said that the dam is well on its way to becoming the base of one very expensive waterfall.

Still, for at least the coming decades, there is little chance that the Glen Canyon or any of the other big dams in the United States will be coming down. The real debate will come if an earthquake or other event causes a catastrophic failure of one of the big dams. If that happens, the likelihood of rebuilding is probably remote. As long as they stand and as long as they still have some room in them to store water, however, they will continue to serve the water storage needs of the people, farms, and factories nearby. They will also provide important flood control for the people living downstream for many

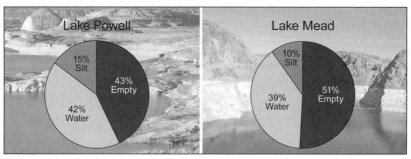

Source: Bird 2003.
NOTE: Lake Mead is even lower today than at the time these estimates were made.
FIGURE 7-3   RESERVOIR STORAGE CAPACITY USED: LAKE POWELL AND LAKE MEAD

years to come. Complicating the picture even more, it is critical to remember that hydroelectric dams also provide one of the cleanest sources of electricity we have in this country. Energy that requires no fossil fuels to produce represents a significant advantage that can't be ignored when global climate change remains a central issue in political discourse.

And to be sure, taking down a dam can also create a whole series of new problems. Whereas sediment from the small Marmot Dam in a relatively remote region of Oregon flowed downstream without causing any problems, when crews removed the Fort Edwards Dam on the Hudson River in 1973, they let loose all kinds of problems. The dam, made of timber and filled with rock, had been built in 1817 and needed to be removed because of safety concerns. After it was removed, all sorts of debris that had been trapped behind the old dam was allowed to flow downstream. Among other problems, the debris jammed up the entrance to an important canal used for shipping. Even worse, many of the PCBs that had been dumped into the river over the previous years and remained trapped in the sediments behind the dam were exposed and began to contaminate the river downstream toward New York City. It's an environmental problem that is still being managed today; in fact, it's one of the largest environmental remediation programs under way in the nation.

In cases where smaller dams may need to stay intact because of the freshwater needs of those nearby, or because of flood control or other reasons, dams may need to be repaired and retrofitted. Here, a whole range of new ideas is being tried around the country. For instance, after the devastating North Dakota Red River flooding in the spring of 1997, the communities in the Grand Forks Region built a dam and a large floodplain that doubles as a greenway with lots of recreational opportunities. The dam itself is equipped with a new design of arched ramps built from large rocks and specially designed vanes so that fish can get by and boats can pass in safety. In other cases, dams may be partially breached, improving the ability for fish to pass but leaving much of the dam for flood control.

In other cases, the dams themselves may just be managed differently. For example, the US Army Corps of Engineers is working with the Nature Conservancy in Kentucky on the Green River. By studying the ecology down the river from the dam, biologists were able to come up with a plan for releasing water from the reservoir at rates and times that would be most beneficial for the fish and mussel species downstream.

In short, there may be as many different futures for dams in this country as there are dams, with each one getting a solution that might range from destruction to just fine-tuning the release of water. As with other scenarios in this book, it's a future that is more complicated, but it's an improvement on the one-size-fits-all approach that so dominated the previous hundred years.

So, the future of water storage in the United States is generally not going to be found in new reservoirs behind new dams. All dams are temporary. Nonetheless, the population of otherwise dry places like California keeps growing. Even as the amount used by each new family drops, as indeed it has been over the last decade or so, the total amount of water needed will grow, and the current storage infrastructure just won't be able to keep up. If new dams aren't the answer, what is? This problem has water officials looking down.

We wrote in the last chapter about the myriad problems of tapping underground aquifers as a new source of water. However, there is a bright and potentially huge future for using other porous aquifers as a place to store water during wet times of the year, and in wetter years, in order to deliver that water when the season or the year is drier. The best example is the biggest underground water storage system in the world: the Kern Water Bank in the San Joaquin Valley of central California.

Built in the deep, sandy soil underneath an area of about 30 square miles outside of Bakersfield, the Kern Water Bank has a capacity to hold about 1 million acre-feet of water, or about 325 billion gallons of water. A number of advantages are inherent in this kind of water storage—most of them related to the fact that this approach mimics more closely the ways in which nature moves and stores water. We mentioned earlier that a significant percentage of the water in Lake Powell is lost every year to evaporation—lost forever to local users into the atmosphere, to be deposited as rainfall somewhere else around the globe. With underground storage, there is almost no water lost to evaporation. If the geologic strata in the region are sound with a solid impermeable layer underneath a large area of sandy soil, then a porous aquifer can hold the water with almost no loss. In cases where the natural geology is not as conducive to underground storage, man-made liners and slurry walls around the perimeter of an aquifer storage project can also help keep the water in place for future use. It obviously takes energy to pump that water back up to the surface when it's needed, but with the Kern example, the pumps are only sunk about 75 feet. Electrical power is still needed, but usually aquifer storage projects are located in relatively shallow sandstone horizons.

Another advantage with the Kern site is that much of the land on top has been turned into a wetland, providing a wide range of benefits. Wetlands, and many of the animals that thrive in them, have reached the brink of extinction in many areas. Trees such as the willow that had all been cut down to make way for farming are now thriving in the land above the aquifer. With the return of the native vegetation, nesting space has been restored for more than forty species of birds that have returned to the area for the first time in generations. This is good news for environmentalists and developers alike, because with the return of endangered species it becomes easier to find other land that can be developed.

The plants on top of the aquifer may use a small portion of the water being stored below, but they simultaneously provide a huge benefit for the water users because of the way in which the plants naturally clean and filter the water. You may remember the *constructed wetlands* we wrote about in chapter 4. Agricultural areas such as Iowa use wetlands to filter runoff from farms. The wetlands vegetation eats up much of the excess fertilizer in the water, which in turn prevents water quality issues as far away as the Gulf of Mexico.

WORLDWIDE TRENDS

This is a trend that will also likely continue, and not just in the United States, where environmental laws and lawsuits are driving so many of the changes toward sustainability. For instance, in southern

Spain, a fish farm has become something of a poster child for the future of fish production. The Veta La Palma farm raises a type of bass in refurbished wetlands that had been drained for agricultural production in the early 1900s by the British. In 1982, a Spanish company bought the land and let it flood again for fish production. The United Nations Food and Agricultural Organization reports that farm-raised fish are now about half of all fish consumed by people, and the percentage is growing. Soon, most fish served as food will not be caught, but raised.

At the Veta La Palma farm, water flows into the wetlands from two rivers polluted with industrial wastes and agricultural runoff, but the water actually flows out cleaner than when it goes in because of the abundant sea life and natural plants that grow in the wetlands. That ecosystem produces a healthy crop of fish that the farmers have chosen to let grow to about 2 pounds each, nearly twice the size of most harvested bass. Interestingly, the farmers report that about 20 percent of the crop is lost each year to birds; more than 250 bird species now inhabit the farm. Just like with the new wetlands outside of Bakersfield, Calif., the birds had all but disappeared but are now back and thriving.

Famed chef Dan Barber, at a TED conference in 2010, talked at length about the farm, and said he was astonished at the notion of a fish-farming operation professing satisfaction that one in five of its fish was lost to predators.[3] Barber said the farmers reported to him that the most important thing was to have a healthy ecosystem: if the birds are healthy, that means the fish are healthy, and production of fish will be sufficient to make the operation sustainable.

The beginnings of a similar trend may also be taking hold in India. Large-scale dam projects are under way or being discussed to be sure, and many environmentalists point to them as a way to produce electricity for a developing nation without burning coal. Another less invasive trend that may catch on is a return to something that was reportedly widespread before colonial rule: *check dams*, perhaps 5 feet or so high, filling unlined reservoirs. Those dams catch some water during rainy seasons and create a small bit of an intermittent wetland. Now, scientists understand that a series of small dams like that don't harm the ecology of the river system in the ways that a large dam does. Such dams also recharge aquifers right in the areas where they can be tapped year-round by local residents.

Even PepsiCo India, which is the largest single food and beverage company in India, announced in 2009 that it was going to start a test project with thirteen check dams surrounding a village in the western Indian state of Maharashtra. The project is tiny given the massive pumping from deep aquifers that PepsiCo and other large corporations do in India at rates many times the recharge rate. The falling water table has driven millions of farmers off the land and into overcrowded cities to look for work. But if this pilot program does well, we can expect to see more small dams going up, perhaps saving the lives and livelihoods of millions of people in rural India.

One other project from yet another dry spot in the world, Australia, may also be the start of a trend that may spread to other parts of the world. Right now in nearly every developed city in the world, an extensive system of storm-water removal is in place. When it rains in every major city around the world that lies near an ocean, the precipitation is quickly and efficiently gathered on roofs, roadways, and parking lots and funneled directly into culverts and waterways that deliver the water to the ocean. The concrete spillways in Los Angeles, nearly always bone dry, have appeared in countless movies with car-chase scenes. All those systems were built to deliver the naturally salt-free storm-water runoff quickly away from the city, even as other water is pumped in to supply the need of inhabitants, often from hundreds of miles away. What sense does this make?

So, scientists in Australia picked an urban area about one-third the size of Manhattan, and instead of letting the water flow out into the ocean, they gathered the water in collectors and pumped it into a nearby field where they created yet another version of a wetland, and the water percolated down into a limestone aquifer. The test was seen as a huge success, and now plans are under way in Australia to expand the program. That's a trend that will undoubtedly spread to other cities near coasts, especially in dry areas. Cities—concrete jungles—are excellent collectors of precipitation, and if it can be delivered to aquifers that can store that water, it can create essentially a newfound source of untapped water.

While systems like this can work well, no single one of them represents a silver-bullet solution to the challenges of more efficient water storage. The capacity to hold 1 million acre-feet of water is great news for those who need the water near the Kern Water Bank, but Lake Powell is more than twenty-five times larger. Put another way, Lake Powell evaporates almost twice as much water in a year as the Kern Water Bank can store. But, what if the Kern Water Bank were combined with a storm-water recovery system, and perhaps a series of check dams? Taken altogether in a sustainably integrated manner, these small, low-intensity projects may be able not only to provide year-round water storage for growing cities, but also to help the overall environment.

That's consistent with other trends we've seen in this book. Before World War II, the solution to water storage needs was dams, dams, and more dams, many of them immense. Today, the solutions are things like aquifer storage on a relatively small and local scale. Those are the projects that will sustain the water needs of populations around the world for many centuries into the future.

❧

[1] See Thomas Fuller, "Countries Blame China, Not Nature, for Water Shortage," *New York Times*, April 1, 2010, http://www.nytimes.com/2010/04/02/world/asia/02drought.html. Also see Marwaan Macan-Markar, "China Flexes Hydropower Muscle," *Daily Mirror* (Sri Lanka) Aug. 30, 2010, http://print.dailymirror.lk/opinion1/19921.html.

[2] Greg Gordon, *Landscape of Desire: Identity and Nature in Utah's Canyon Country*. Utah State University Press, 2003.

[3] "TED" stands for Technology, Entertainment, Design. See Dan Barber, "How I Fell in Love With a Fish." TED Blog. March 10, 2010. http://blog.ted.com/2010/03/10/how_i_fell_in_l/.

CHAPTER 8

THE FUTURE OF
WATER UTILITIES

Two days before Christmas 2008, millions of people tuned into a morning news show on every local or cable news channel and watched live video of a dramatic helicopter rescue of nine people stranded in a flood on a road just outside Washington, D.C. "There were boulders coming down the road the size of laundry baskets," a local firefighter told the *Washington Post*. "It felt like whitewater rapids."[1]

But it was no flood. A 66-inch prestressed-concrete underground water main burst near the busy road on a morning when temperatures had dropped below freezing. After nearly an hour in the torrents of water and debris, rescue workers saved all those trapped by the rushing water, using helicopters, boats, and fire trucks. Nobody was seriously hurt. The repairs closed a busy commuter artery until New Year's Day, water was cut off to thousands over the Christmas holiday, and utility officials told thousands more not to use the water in baby formula or to do laundry because sediment in the water might discolor their clothes—while insisting that the water was still safe for adults to drink.

While this burst pipe was certainly the most dramatic problem of the year for the Washington Suburban Sanitary Commission (WSSC), it wasn't the only problem the utility had with a water main. The utility later reported that the break was one of 1,709 for that year, a number that was actually down from the 2,129 breaks in water mains recorded in the previous year of 2007.[2] That was not the number nationwide; that was the number for this one utility district.

We'll begin with a look at the future of the largest utilities in the United States before moving to the medium-size and smaller utilities here, and then we'll look at utilities around the world. The USEPA categorizes utilities as large if they serve more than a hundred thousand people, and currently about six hundred such utilities exist in the United States. The differences among those utilities are many; most are owned and managed by regional governmental authorities, but a few are privately owned (more on that in the next chapter); some are part of a broader city or county government and some are their own separate entities; some are part of a broader utility agency that may manage wastewater or power in addition to drinking water; some have all their own raw-water sources locked up; and some purchase water from others before delivering it to customers. Geography, history, demographics—and particularly politics—tell the story of each utility district.

LARGE UTILITIES

While the Washington Suburban Sanitary Commission is unique, it's also quite typical. It was founded early in the twentieth century and has hundreds of miles of water mains that are approaching the end of their useful life span. As the nation learned during that December day in 2008, even pipes that are still technically within their expected life span may not be up to the task. The break released 150,000 gallons of water per minute, gushing so much water that 6-foot chunks of asphalt swept down the road. It took crews 5 hours to shut off the main.

The problem, as it turned out, was not the 43-year-old pipe itself, but the nature of its installation. According to a report later issued by the utility, the pipe was resting on rocks instead of on a smooth bed, and this was what caused the problem. The pipe had held up since

1965, but on that frigid morning, it ruptured. The repair and cleanup cost the utility $1.67 million. That's not a huge amount of money in the big picture of a large utility like the Washington Suburban Sanitary Commission, which serves 1.8 million people. But consider that the WSSC has 5,500 miles of water mains, and at least 1 in 4 of those miles is more than 50 years old.

At the time of the break, the WSSC had a budget to replace only about 25 miles of pipe each year, meaning that it would take about two hundred twenty years to replace all of the mains using the existing budget. And the simple fact is that water pipes—concrete, ductile iron, or even steel—don't last for two hundred twenty years. And therein lies the essence of the problem of our dilapidated water infrastructure problems in this country. Our underground water distribution infrastructure may be out of sight and out of mind for most people, but we are not maintaining it at anywhere near the rate we should. Hence, it is gradually crumbling. This, in a nutshell, is one of the great challenges of the US water utility industry: to convince ratepayers and the general public that we need to be paying more for our water so that we have the money to sustainably manage our infrastructure into the future. It's a critical social and economic challenge facing the country, but it goes largely unnoticed.

Earlier in 2008, D.C.-area papers reported that the commissioners of the WSSC had rejected a new fee of $20 a month per customer. That fee would have funded an accelerated schedule of pipe replacement, but the commissioners said the flat fee was unfair to low-income customers. The commissioners did not, however, change the fee to a sliding scale. They decided to ask for money from the federal stimulus package instead; however, they didn't get any. The WSSC did get some stimulus money for some long overdue sewer work, but none for water mains. Hence, the status at that utility today remains more or less the same as it was the morning of that water-main rupture and dramatic rescue seen around the world. And this is typical of the water utility business; once the emergency passes, public awareness and concern tends to fade away as well, but the underlying problem remains and continues to grow more critical.

So, what is the future of large water utilities, especially given the aging infrastructure they rely on? We may not know for sure, but everyone agrees that there is a problem, and that the problem is huge. The most recent estimate from the USEPA is that we'll need to spend about $335 billion over the next twenty years to maintain and extend our water treatment and distribution infrastructure. These six hundred or so large systems will represent about a third of that total, or around $115 billion. On a simple average, that means each large utility has a need of nearly $200 million. Put another way, we need about $1,000 for every man, woman, and child in the United States to address this challenge—and that's on top of the amount we already pay in taxes, water, and tap fees. Nobody knows where this money is going to come from.

And all of this is happening at the exact same time that utilities face another puzzling paradox: water utilities, particularly in the more arid western part of the country, have been focused for years on trying to devise and implement broad-scale conservation programs to encourage their customers to use less water. At the same time, utility managers are encouraged to run their agencies more like businesses. The less water the customers use, the lower the incoming revenues of the utility. How do you convince economically strapped consumers to pay more per gallon of water even as they struggle to use fewer total gallons? As we mentioned earlier, the Southern Nevada Water Authority has even paid its customers to take out turf grass to cut water consumption. As Director Pat Mulroy has quipped, "We are the only industry in the world that pays our customers cash to convince them not to buy our product!"[3]

These paradoxes and questions lie at the heart of the challenges that utilities will face over the coming decades. Because of the high cost of the infrastructure and the prohibitive costs of moving water over long distances, there's not much chance that we will see radical changes in the structure of the water utility business. The telecommunications and electrical utility industries *have* seen a radical competitive change over the past few decades. However, water is infinitely heavier than electrons, and you can't easily wheel (or move) water around the way you can electricity. Hence, the physical water and wastewater infrastructure will continue to be managed as some sort of monopoly service, probably forever. Public or private, water utilities will almost certainly continue to be what the economists call *natural monopolies*—in other words, it just doesn't make sense for a neighborhood to have two or three sets of water pipes and competing water providers.

So, these regulated and publicly managed water utilities are encouraged to operate more like businesses, but they have little ability, or political willpower, to bring in greater revenues at the same time that their deficiencies or needs are bursting forth for all to see on the morning news.

What will come next for the water utilities? The future of the water utility business depends in many ways not so much on science or economics, as was the case in our previous chapters, but on public perception and the government. Water utilities basically do two things: they treat water, and then they distribute it. And all utilities—public and private water utilities, large and small—are highly regulated, particularly in terms of the treatment part of that overall job description. The USEPA sets precise standards for the quality of water as it leaves the treatment plants; interestingly, though, the USEPA does not regulate the water once it gets into the pipes and on its way to our homes and businesses (more on that in a minute). A vast body of water regulations is on the books, but let's first take a look at those regulations that concern the safety of drinking water.

MAKING WATER SAFE TO DRINK

Right now, the USEPA mandates that water utilities test water for about a hundred contaminants in six categories: microorganisms, disinfectants, disinfection by-products, inorganic chemicals, organic chemicals, and radionuclides. Acceptable levels have been established for some of these contaminants, while for others there is zero tolerance. The USEPA has changed the requirements for some of these contaminants, but the list has only grown since the Safe Drinking Water Act was passed in 1974 as shown in Figure 8-1 on page 140. It's almost certain that this list will continue to expand in the future.

In 2009, the USEPA published its third Contaminant Candidate List, a list that comes out every six years and identifies additional compounds and chemicals that may have a harmful effect on human health or the environment. The list currently has almost one hundred additional compounds that may be regulated or banned in the future—and environmental and health advocates say that it should be far more comprehensive.

This candidate list includes a whole new range of contaminants: nine different hormone compounds, numerous medical drugs, and a whole range of pharmaceutical products. For example, if you've had about any kind of infection in recent years, you've probably been prescribed erythromycin. Health officials are concerned that too much exposure to antibiotics like erythromycin tends to make the medication less effective, and increases the likelihood of so-called superbugs that are increasingly resistant to traditional antibiotics. Groups such as Keep Antibiotics Working say that, even at very low levels, the constant exposure to antibiotics can lead to greater resistance. That concern may mean that the USEPA will instruct water providers to treat water so as to remove these antibiotics—and that's why erythromycin is on the candidate list.

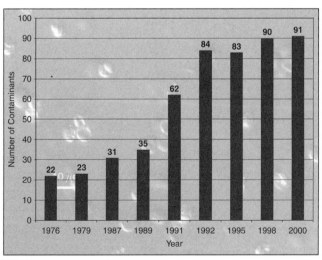

Source: USEPA 2001.

FIGURE 8-1 NUMBER OF REGULATED CONTAMINANTS FROM 1976 THROUGH 2000.

Similar to many other pharmaceuticals, antibiotics not absorbed by the body get flushed down the toilet, move into the wastewater treatment plant, and eventually show up downstream and enter another utility's raw water intake. That's why fourteen pharmaceuticals are on the Contaminant Candidate List. The big problem from the utility's perspective is that from a treatment perspective these pharmaceutical compounds are quite different from microorganisms and other chemical contaminants. Filtering, settling, and treating with chlorine works fine for most of the currently regulated contaminants but will do nothing to remove very low-level pharmaceutical compounds from water.

In 2008, the Associated Press (AP) asked many of the largest water providers in the United States about any tests they had performed to look for pharmaceuticals in the water. Of the twenty-four that provided information to the AP, only three found no evidence of pharmaceuticals.[4] The USEPA doesn't require utilities to publish the results of these kinds of nonrequired tests—and the analytical science is so new and the amounts are so small that most water districts or companies aren't anxious to. It's also not at all clear how much of a health hazard is actually involved. The amount of most of these pharmaceuticals in the water is very tiny, registering in the parts per billion, or even parts per trillion. (Of course, they *are* intended to work at very low doses.) In fact, it's only quite recently that we've even had microscopes and instruments sophisticated enough to see these compounds in the water. The potential concern comes not so much because of the concentration level, but because we consume lots of water every day, and few thorough and comprehensive long-term studies have been made of the cumulative health effects over time of continuous exposure to these various compounds.

That AP series, combined with similar press reports, widespread information published on blogs, and hearings in front of the US Congress about water safety may be in part what will prompt the USEPA to add various pharmaceuticals to the banned list. Here's where the issue takes on a political twist. Any parties lobbying to keep such pharmaceuticals or other compounds off the list find themselves in the difficult position of appearing to advocate standards that go against public opinion. It just doesn't sound good to go in front of Congress and say that you think it's fine that there are some pharmaceutical remnants in drinking water. For this reason, it seems likely that the range of regulated drinking water contaminants will only continue to expand in the future.

So if water utilities *can't* argue against treating for a new range of contaminants, they certainly *can* argue that they should be able to increase taxes or charge more to treat those contaminants, right? Well, this of course is another paradox for utilities that need to maintain and replace not only aging pipes, but also aging and ineffective treatment plant facilities. Like many of the pipes, water treatment facilities are also often more than fifty years old, and without major upgrade work, many of them are coming to the end of their expected life span. So in addition to needing to repair or replace treatment plants to keep up with current regulations, the utilities may need entirely new treatment facilities to keep up with where the regulations are going.

Water utilities are also likely soon to face another whole new area of regulation, in terms of monitoring and testing requirements. While utilities currently conduct all of these sophisticated tests on about a hundred different compounds, it's critical to remember exactly where that test is actually done: at the discharge point in the water treatment plant, right before the water is released into the distribution system. Essentially no further monitoring or testing of that water takes place.

Considering that many of our pipes are leaking badly, it stands to reason that nasty things can also find their way *into* water pipes. Pipes can corrode or experience biological buildup, which may not be particularly good for the drinking water. Not only that, but some of our pipes are still made out of lead, which can leach out into the water, and which is a well-known human-health problem. Finally, concerns have recently been raised about the potential for terrorists to actually attack the water system, or to inject poisons into the public drinking water systems. So, the bottom line is, even if our drinking water is pretty safe and clean when it comes out of the treatment plant, it may not necessarily be quite so clean when it comes out of our tap. For that reason, it seems likely that utilities will face a plethora of new distribution system monitoring requirements in the future, adding another dimension to the complexity of their job.

WASTEWATER TREATMENT

Speaking of other dimensions, we also need to discuss a topic that we've only touched on so far, but that is also under the umbrella of the utility industry. It's time to dive a bit deeper, so to speak, into the septic pool and talk about wastewater treatment. Indeed, in some ways, the American public pays even less attention to wastewater issues than it does to drinking water issues. Industry estimates hold that the wastewater industry in the United States is only slightly smaller than the drinking water industry; as a country, we pay roughly the same amount to dispose of our wastewater as we do to acquire our drinking water. And in both areas, the treatment is fairly similar: both take in water that is impure and both discharge treated water after removing solids, organic and inorganic compounds, and so on. And both sectors have similar unfunded needs. The Congressional Budget Office reported in 2002 (one of the most recent comparative analyses) that the unfunded needs for drinking water infrastructure are $20.1 billion per year, while the similar figure for wastewater needs is $20.9 billion per year.[5]

Both areas are regulated by the USEPA, though under the guidance of different acts of Congress and from within different divisions in the agency. And to complicate things further, it is typically the individual states that are charged with monitoring and enforcement of these laws, and actual enforcement levels can be highly variable. The Safe Drinking Water Act does what its name implies, and it is this expanding body of legislation that will probably eventually regulate what types of pharmaceuticals are acceptable at the tap. Another arm of the USEPA regulates the quality of treated wastewaters that can be released back into lakes and rivers. The Clean Water Act, like the Safe Drinking Water Act, does not ban pharmaceuticals from entering the waterways right now, but this situation will probably not last for long, either. For example, scientific studies have recently reported on hermaphroditic organisms living in creeks and rivers below wastewater treatment facilities in various parts of the United States, Europe, and Japan. These fish and amphibians have both male and female sex organs, and as you might imagine, they are changing and disrupting the local ecology.

The reason for this sexual confusion in the fish is thought to somehow result from human activities that contribute unnatural chemicals into the surface waters—including birth control hormones, medicines used by the sick among us, ranchers feeding growth hormones to cattle, the shampoos and health and beauty products that we all use, and various other things. The problem is widespread and in many areas is beginning to reach epidemic proportions. In the Potomac River, three of four male bass reportedly have immature female eggs in their testes. In one tributary of the Potomac, the Shenandoah River, every single male bass contained female eggs in a recent test.

The USEPA is taking steps to encourage health care facilities to properly dispose of all medications and not to flush them down the toilet as they've been encouraged to do in the past. But people who use medications cannot be stopped from using the toilet. Urine is the means by which we rid our bodies of unneeded medications (among other wastes), and typically as we grow older, we use more medications. With an aging population in the United States, and with growing use of a whole new range of pharmaceuticals, this is an issue that is going to get worse and that wastewater facilities are going to have to face. And as more and more environmental advocates warn, pharmaceuticals in natural waters may be sort of like the traditional canary in the coal mine: a warning signal to the human population that we are getting into a dangerous area.

Just as is the case with drinking water, those facilities that treat wastewater are aging; many of them have already zoomed past their planned life span. As we said, the estimate for unfunded but needed repairs and upgrades to water treatment facilities is approximately the same as it is for drinking water: a number so large that is essentially meaningless, because the political will to write such a check does not exist. And just as with drinking water plants, the list of contaminants the wastewater treatment plants need to reduce or eliminate is almost certainly going to be getting bigger. And many of these additional compounds are going to be low-level and extremely hard-to-treat items like pharmaceuticals.

Furthermore, even if the money *can* be raised or *does* exist to build new wastewater plants, treatment facilities that process domestic wastewater are dramatically less popular from a public perspective than drinking water plants, which aren't all that popular in the first place. The odors and by-products associated with the treatment of a wastewater stream mean that a not-in-my-backyard attitude will pervade any planning and siting discussions for new wastewater treatment plants. In turn, wastewater plants will be located farther and farther away, increasing the costs of running new pipes into those plants.

INNOVATIONS IN WASTEWATER TREATMENT

So, what else can we expect in terms of the future of wastewater plants? Because of the expense of building new plants, and because of a growing trend to mimic natural water treatment systems like wetlands, it's likely that we'll see a trend toward new and innovative systems that don't look like they have much to do with wastewater but really do. For instance, New York City started a plan in 2010 in which it will pay building owners to install rooftop gardens. At first blush it doesn't sound like a tomato garden on top of a condo has much to do with wastewater, but in fact it does.

In New York, the storm sewer system is built in with the sanitary sewer system; that is, the rainwater rushing down the streets and into the gutters eventually flows into the sewer system that contains all of the domestic wastewater. As you might imagine, faced with a big rainstorm, the wastewater system can get overwhelmed pretty quickly. This is actually a common situation in most cities in the country, referred to as combined sewer overflow. In the early 1900s, the city of New York installed a particular kind of valve in its combined sewage system. Whenever a big rain hits New York City, those valves kick in to protect the aging treatment plants, and the combined storm waters and sewage run straight in to the Hudson River or the sea with no treatment at all. To fix New York's wastewater systems, it would cost the city an estimated $36 billion, and that's not likely to happen in a huge and sprawling city with hundreds of competing fiscal requirements.

So, back to the tomatoes. That's why the city has kicked off this plan to install rooftop gardens and a whole array of other initiatives to soak up and use rainwater before it rushes into the sewer system. Compared to at least $36 billion for a new system, this plan seems like a bargain at only $1.5 billion over the next twenty years. This concept is increasingly known as *green infrastructure*: working against the vast concrete surfaces of a modern city with systems that can make actual productive use of that rainfall, as well as it keep it from flooding the sewer systems and wastewater treatment plants. It's something that New York and many other cities can do to forestall legal enforcement action by the USEPA and ultimately to avoid sending raw sewage into the natural waterways. This technique is popularly referred to as "spread it out, slow it down, and soak it in."

We wrote in the last chapter about a project in Australia trying to capture storm runoff and inject the water into underground aquifers. Australia is trying that because of chronic drought conditions, and arid Western cities like San Diego are thinking about doing the same thing. But don't be surprised if we see more and more efforts like this even in Atlantic Seaboard cities that get plenty of rainfall. Capturing rain and snow runoff can represent a double-whammy solution in many of these circumstances: not only does it prevent overloading of the sewer system, but it also can help recharge depleted groundwater in places where seawater is intruding into and contaminating the aquifers. Green infrastructure is a trend that is becoming well established now and will be a big part of the future of water.

In general, the trends we've written about in the chapters on water use in the home and industry should also have a positive impact on the wastewater treatment picture. Even with a growing population, many wastewater utility districts should be able to extend the life of some existing plants, because declining primary-water consumption also implies that less wastewater volume will be flowing into their systems: less water in, less water out. Plants will clearly need to upgrade because of the environmental concerns and regulations mentioned previously, but the actual gallons processed should remain low enough to forestall the need to build many new plants, at least in the United States. What happens, however, if the amount of water processed gets too low?

Ed Means, a highly regarded water industry veteran, points out that most wastewater plants rely on a gravity-fed system to constantly move raw sewage along to get to the treatment facilities.[6] If water use falls dramatically in our homes and businesses, there just may not be enough water to move the solid waste along, creating potentially dangerous buildups. Water use hasn't fallen anywhere near this far yet, so this isn't a problem that's endangered any people or facilities so far, but this is the kind of paradoxical outcome that we seem to often see in this business: the creation of a new problem that results from a well-meaning effort to solve some other problem. We can expect to see other examples of this law of unintended consequences in the future.

Wastewater treatment plants have their job to do, and in general consume money, energy, chemicals, and other resources in cleaning up dirty water before returning it to the environment. But they also produce other by-products as an output. In recent years, many plants have taken the solid waste by-products left over from the treatment process and turned them into fertilizers, saving stress on landfills and creating an environmentally friendly soil additive. Typically, federal or state regulations prevent these *biosolids* from being used to fertilize human food sources, but they can be used on Christmas tree farms, wetlands, and other nonfood agricultural areas. Some municipalities use biosolids to fertilize flowers and other plants installed near highways or in parks. That's a trend that will certainly grow in importance, especially as agricultural producers search for green alternatives to petrochemical by-products, the current source of most commercial fertilizers.

Wastewater plants will contribute much more in the future. All of that stuff that the plants remove from the wastewaters is probably in there because someone was using it for a worthwhile purpose to begin with. For example, a common constituent of most wastewaters is phosphate. Phosphates are commonly used in agricultural fertilizers and provide critical nutrients for biological life. Various forms of phosphate are also commonly found in the dishwashing and laundry detergents that we all use every day. However, if uncontrolled, excessive phosphates can also lead to massive over-fertilization and growth in natural waters, or *eutrophication*. Traditionally, the phosphate content of wastewater has always been thought of as a serious waste problem.

At the same time, it turns out that the world is rapidly running out of natural phosphate deposits. The phosphate mines of Saskatchewan, Morocco, and Florida are rapidly depleting, and phosphates basically have no substitute in terms of the role they play in agricultural fertilizer applications. In other words, if we run out of phosphate, we're going to have trouble producing enough food. In fact, Isaac Asimov, the well-known science fiction writer, once even defined phosphorus as "life's bottleneck." So, today, a lot of companies and entrepreneurial technology developers out there are looking at ways to kill two birds with one stone: getting the harmful phosphate out of wastewater, and then gathering and concentrating it for reuse in critical agricultural applications.

Or take another example. Right now, about two hundred homes in the British town of Didcot, Oxfordshire, have a gas line to their home that is being fed by methane gas captured in a nearby wastewater treatment plant. Officials there estimate that if each of the nine thousand six hundred plants in the UK installed a gas-capture device, together they could provide gas to heat two hundred thousand homes.

Nonetheless, the single most important thing that wastewater plants produce is water. As we discussed in chapter 6, the traditional sources of raw water available to be transformed into drinking water by treatment plants are being depleted. Aquifers are either drying up or becoming fouled with seawater. No new dams are coming online. Desalination is expensive and doesn't really work if you're located inland. Hence, *direct* connections from wastewater plant outlets to drinking water plant inlets will almost certainly be more common in the future here in the United States, just as they are already common in several water-short places around the world.

As we've said, indirect reuse of wastewater has been with us since ancient history—we're still using the same water molecules that the dinosaurs and ancient humans used. But we aren't really that far away from the scenario of more direct reuse right now. Wastewater plants discharge into the same rivers that are the sources of primary water downstream. Water in the Mississippi drawn out in New Orleans has obviously been used and reused over and over again, in the process of draining parts of thirty-two states and two Canadian provinces. Right now, we have the psychological comfort of knowing that the water has been mixed in and somehow combined with the regular flow of the river. We derive the same apparent comfort when wastewater flows are discharged into underground aquifers at one point and withdrawn as primary raw-water sources a few miles away. It brings to mind the old saying that "the solution to pollution is dilution." And we seem intuitively confident and comfortable that Mother Nature will somehow clean those waters before we extract them to use again. It's a comfort we may not be able to hold onto much longer.

These are just a few examples, but in the future, wastewater treatment plants will be increasingly thought of as resource and energy production plants. In other words, wastewater isn't really a waste. In fact, it's no such thing: it's a source of clean water, as well as a source of various other types of depleting and critical natural resources. In the future, we'll make less of a distinction between *water* and *wastewater*—everything will just be water.

A LOOK AT THE SMALLER UTILITIES

We've focused most of this chapter on challenges and issues affecting the largest water utilities in the United States, and most of these challenges will have a similar impact on the smaller utilities, too. But a lot of these smaller utilities are going to be increasingly hard pressed to face effectively the plethora of future challenges—scarcer water resources, increasing costs, more sophisticated technologies, and tougher regulatory requirements. As we'll see, one trend seems

TABLE 8-1 DISTRIBUTION OF U.S. WATER UTILITIES BY SIZE
AND POPULATION SERVED

System Size	Number of Water Utilities	Percent of Water Utilities	Population Served	Percent of Population Served
Large Water Utility Systems (serving over 100,000 people)	584	1	128.6	45
Medium Water Utility Systems (serving 3,300–100,000 people)	8,749	17	130.7	46
Small Water Utility Systems (serving fewer than 3,300 people)	41,748	82	24.1	9
Total	51,081	100%	283.4	100%

Note: The remaining population not accounted for here is served by individual wells or non-utility systems.
Source: USEPA 2009a.

almost certain: consolidation among the ranks of these more than fifty thousand smaller and medium-size utilities (see Table 8-1 above) in the United States. Some mergers have already occurred between small and medium-size water utilities, with many more sure to come.

Two factors make this prediction something of a certainty: complexity and efficiency. Large utilities—like New York, Los Angeles, or Denver—can afford to hire experts to study alternative treatment technologies, to analyze complex overlapping regulations, or to evaluate different approaches to solving pharmaceutical contamination. The small utility simply can't afford any staff that aren't directly involved in delivering water, billing customers, or operating the plant. Larger utilities, public or private, can invest in improvements because the cost is spread over a much broader base. Despite the political challenges of combining or merging public agencies, many industry experts believe that we will see widespread consolidation within the utility business; the only alternative is widespread noncompliance with existing regulations.

In addition, water utilities are becoming more central to many of the closely watched issues of our time. In many communities, especially in the West, environmental activists are using water issues to limit or stop new housing and commercial development. Activists use any law or federal rule they can to stop urban sprawl, particularly if the development can't be responsibly supplied with sufficient water. Sometimes it's not environmental activists but the water utilities themselves simply making the point that they don't have enough water for the people they already serve, so it's not responsible for them to add new customers. Another rub is that in many districts, attempts to raise rates on existing users have been politically unacceptable, so instead they increase the price for tap or connection fees on new homes. For many water utilities, it is only these new tap fees that have allowed them to balance their budgets. But this can also be a no-win sort of paradox. A decision to cut off new tap fees as a source of revenue can cripple the budget, but allowing continuous growth of new taps and new customers can cripple the ability to serve everyone in the future. It's just another one of the complex and difficult trade-offs facing all water utilities, but especially the small and medium-size ones.

And in smaller but highly visible ways, water utilities are being held under the microscope of public attention once reserved only for legislatures and city councils. Communities are demanding, for example, that utilities purchase hybrid or flex-fuel fleet vehicles, or that they print bills on recycled paper or deliver them over the Internet. Large districts can handle such requests without a significant hit to the overall budget. Small- and medium-size utilities have a harder time.

Also, innovative new technologies will help to make large utilities more efficient, while smaller utilities may not have the cash available to keep up. Right now, larger water utilities around the country are installing *smart meters*. These meters measure the amount of water flowing into a house, for example, around the clock, and then they report on that usage over wireless networks. With precise measurements and user history, the meter itself can alert both the utility and the customer to a possible leak on the customer's side of the meter. With accompanying sensors, utilities can zero in on leaks much more quickly and accurately. And in turn, crews can get out to the leak while it is still just a leak and before it becomes a full rupture, avoiding explosive disruptions and huge costs. Smaller districts are going to have increasing challenges spending the kind of up-front money needed to make efficiency improvements like that.

And remember the paradox we wrote about? Water utilities are under tremendous pressure to operate more like a business, yet they also have an absolute need to reduce the amount of water flowing to each home. Small districts are run by neighbors and community leaders. The stark reality is that it's hard for a small community leader to tell his or her neighbors that they need to use half as much water and pay twice as much each month. It's hard for big utilities, too. Remember that just before that water main break in suburban Washington D.C., the commissioners in that district voted down a fee of only $20 a month, and that district is the eighth largest in the country. The additional $20-per-month fee was only enough help the district cover its expenses. Even so, the district didn't have the political will to put that fee in place. In smaller districts, the need is just as great, but the political pressure to keep prices low is even greater. As harsh as it sounds, the only way that small communities may be willing to pay a price that is closer to the actual cost is if the people determining that price are nameless, faceless, and far away. That also helps to explain why smaller districts will be joining together and joining larger water utilities.

How will it happen? Consider again the issue of pharmaceuticals in water. If the USEPA makes elimination of those contaminants a requirement or places some other regulatory burden on drinking water providers, many small districts will simply have to concede that they don't have the financial wherewithal or the operational ability to comply with the new regulations. A cascade of mergers and combinations will almost certainly follow.

WATER UTILITIES AROUND THE WORLD

Water utility organizations in the United States are not all cut from the same cloth. Some are public; others are privately owned by investors. Some are massive and serve millions of people, while thousands of others are so small they each have only a few thousand customers. But as different as they may seem to us, they appear homogeneous when compared to the utility systems currently being employed in places like Beijing, Brussels, or Buenos Aires.

Half the water in the municipal system in Beijing, for instance, is lost to leaks or theft. This is in a city that is in a permanent state of water shortage because of exploding population and a relatively dry climate. The lack of water is a huge reason why many of the biggest dam projects in the world are being built now in China, even though the dams are hundreds or even thousands of miles away from the people who need the water. While fixing the leaks in the municipal system would be expensive, it is still far less costly in financial and environmental terms in comparison with building the biggest water diversion project in the history of the world. That's not the path China has chosen, however.

Many encouraging success stories are out there. The small North African country of Tunisia has been successful in reclaiming 30 percent of its wastewater and has plans to reclaim up to 60 percent in the coming years. That water is used and reclaimed in the northern part of the country near the population centers, then piped south to the edge of the Sahara Desert, where the water is used to irrigate vast agricultural lands. That water has helped the country build a thriving export business of food that is sent to Europe.

Or take the example of Singapore, the small island nation at the tip of the Malay Peninsula. Built on one main island that's a bit smaller than all of New York City and containing no natural lakes at all, for years Singapore has had to import water from Malaysia. The country decided that it wanted to be water independent, and has recently taken a number of steps to accomplish that—and in the process Singapore has become one of the real global leaders in water technology and conservation. For example, it collects storm-water runoff in a gradational system; it separates out the waters that would be hardest to treat and uses that for irrigation. The cleaner water gets treated and used for drinking water. It's used carefully, however. Singapore launched a massive public relations campaign that really worked: per capita consumption dropped dramatically in the first decade of this century. Singapore also recycles a lot of its wastewater. The Ministry of the Environment and Water Resources branded the recycled product *NEWater*; it is created at a series of five treatment facilities located around the small country. The water is used in homes, bottled and sold at retail, and delivered to manufacturers needing highly purified water. Of all the sources of water used in Singapore, it is actually the most pure.

While Singapore is interesting technologically, the most important lesson other utilities around the world may learn is that perceptions matter. By making the case with great force using all possible avenues of reaching out to the public, the Singaporean utility has been able to convince the population that water matters. Once that public education battle is won and people begin to develop a better understanding of the real challenges of water scarcity, it's much easier to convince them to pay more for water, to use recycled wastewater, and to generally use less water around the home.

That trend, perhaps more than any other, is what may define the water utility of the future. Technological advances certainly matter, but what matters most is the ways in which humans use and pay for water. Right now, most people in most countries don't value water very highly, even though they know that without it they would die. The most important job utilities around the world may have in the coming decades is convincing people that water is valuable—and that it is reasonable to pay more for this luxury than the bargain prices we have traditionally taken for granted.

There are many other shining examples of water utilities taking smart steps and utilizing better science and improved management practices to improve their service. Those steps are not only helping the utilities to guarantee that they will be able to provide water for many years to come, but also allowing them to do so without overly taxing nonrenewable sources of water or energy.

Let's be clear, however, about the limits of what utilities can do. Improved management, new technology, and more dollars can help, but they will never solve all the challenges. As the old saying goes, "God gave us free water, but he forgot to lay the pipes." If Beijing, for example, fixed all those leaks, it still would be facing the fact that China has 20 percent of the world's population and just 7 percent of the world's freshwater. Australia can encourage people to take two-minute showers, down from the four minutes it urges now, and it still won't have enough water. Utilities in both of those countries will need to work tirelessly just to maintain an acceptable amount of water for their people. Water scarcity will simply be an increasingly critical issue in many parts of the world.

We should also note here that a huge portion of the world is not served by any water or wastewater utility at all. About one person in six does not have access to clean drinking water at all, and perhaps two people in six do not have adequate wastewater facilities to safely flush away their wastes. For all of the issues they face, the drinking water and wastewater utilities of the world do a remarkable job of keeping people healthy. Those not served by water utilities face grim prospects in simply surviving from year to year.

The good news is that nobody is in a better position to deliver that message than the water utilities. In a study published in 2010 by the Water Research Foundation, researchers asked more than six thousand people about their attitudes toward water prices and more. The people surveyed said they trusted water utility managers as the single most credible source of information about water. Those surveyed said they did not trust elected officials, the media, or those selling water-saving devices at retail stores. Water utilities have perhaps been too timid in the past in making the case for conservation, reuse, and greater infrastructure investment—and for pointing out that cheap, clean drinking water is one of the great luxuries and economic bargains of all time. Armed with confidence in this underlying strength, and the fact that they are the right people to deliver the message, perhaps water utility managers will be better able to go out and make this case to the public in the future. It is only with higher prices and increasing capital investments that we will all be able to continue to have safe drinking water. Not only that, but when we turn on the TV, we won't have to watch people being rescued by helicopter from whitewater rapids created by a burst water main.

Ↄ

[1] Dan Morse and Katherine Shaver. "Water Main Break Forces Dramatic Rescue of Nine." *Washington Post,* Dec. 24, 2008. http://www.washingtonpost.com/wp-dyn/content/article/2008/12/23/AR2008122302853.html.

[2] Washington Suburban Sanitary Commission. "2008 Continues Water Main Break, Leak Trend." Jan. 6, 2009. http://www.wsscwater.com/home/jsp/misc/genericNews.faces?pgurl=/Communication/NewsRelease/2009/2009-01-06.html.

[3] Patricia Mulroy, Keynote Presentation to the First Annual American Water Summit Conference, Nov. 3, 2010, Washington, D.C. http://fora.tv/2010/11/03/American_Water_Summit_Speech_by_Patricia_Mulroy#Colorado_River_Drought_Spurs_Seven_State_Pact.

[4] Associated Press, "Drugs in the Drinking Water." An AP Investigation: Pharmaceuticals Found in Drinking Water. 2008. http://hosted.ap.org/specials/interactives/pharmawater_site/.

[5] Congressional Budget Office, *Future Investment in Drinking Water and Wastewater Infrastructure: A CBO Study.* November 2002, page ix. http://www.cbo.gov/doc.cfm?index=3983&type=0

[6] Phone interview with Scott Yates, August 2010.

CHAPTER 9

THE FUTURE OF THE WATER BUSINESS

To understand where the future of the water business is headed, we'll first look at the current state of the water business. First, make no mistake about it: water is big business. Very big. It's impossible to define exactly, but taken as a whole it's probably the world's third largest industry, or even second if you lump the oil and gas industries together with the electrical power industry and just call it one big energy industry.

THE CURRENT WATER INDUSTRY

It's fruitless to even try to categorize or count all the companies in the water business. Just a few of the larger water industry sectors would include water treatment equipment makers; pump, valve, and pipe manufacturers; water engineers and plant designers; water testing laboratories; monitoring equipment manufacturers; contractors that dig trenches, install pipes, and pave streets; and manufacturers of treatment chemicals. The list goes on and on, and it continues to expand as water problems become more and more severe. Sometimes the only thing that these widely diverse companies and people have in common is simply that they each play some role in finding, delivering, treating, or distributing drinking water or wastewater. Table 9-1 shows an estimate of the breakdown of just the US water industry.

TABLE 9-1 THE US WATER INDUSTRY

Segment	2010 Revenues ($ millions)	2010 Growth Rate Percent
Water Treatment Equipment	10,900	4.6
Delivery and Infrastructure Equipment	12,230	2.0
Water Treatment Chemicals	4,340	3.0
Contract Operations	2,930	2.6
Consulting and Engineering	9,160	2.9
Maintenance Services	2,080	2.5
Instrumentation and Monitoring	1,180	5.0
Analytical Services	920	2.0
Water Utility Revenues	46,910	6.5
Wastewater Utility Revenues	42,050	3.6
Total	132,700	4.4%

Source: Adapted from EBI 2010. Used with permission.

Most experts now place the size of this commercial water market at somewhere between $500 billion and $600 billion per year worldwide. And while people may debate the absolute size of the business, everyone agrees that it's growing at a good clip, and that it will almost certainly continue growing for a long time into the future; our water challenges and shortages aren't going away any time soon. Also, a lot more folks are planning to get into the water business, not just from various sectors of industry, but also from the financial and investment communities. With all the attention the media is giving the impending water crisis, a lot of private firms are getting dollar signs in their eyes. They're jumping over one another to get into the water business, to help solve some aspect of the water challenge, to "do well by doing good."

From a longer-term historical perspective, this is nothing new; private companies have always played a major role in the water industry. Just as water issues are getting more public attention, the water industry itself is coalescing into more of a distinct business sector, and it's getting a lot more attention from investors, analysts, and the popular media. Read any business and investment magazine these days, and you are likely to see an article about some angle of investing in water or a specific water related company. All of those goods and services the industry provides will continue to be more critical in the future, and as our problems become more severe, a vast array of new business opportunities and companies will emerge. In fact, that process is already under way and booming.

THE RUSH TO INVEST

Investors have flocked to the water industry in droves over the past few years, and we don't just mean individuals trying to find a good stock to buy. Investment interest in the water industry has come from across the spectrum: booming private equity firms, mutual funds, foreign sovereign funds, and the whole range of large industrial companies like General Electric, Siemens, 3M, Home Depot, and so on. Private industry obviously sees a huge opportunity in water; for years now, they've been studying and evaluating all of the same trends we've been writing about in this book. The companies hoping to carve out a niche in the water industry understand that water prices will be inexorably going up, and that more and more money will be spent on every acre-foot, gallon, and drop of water. A whole array of (sometimes unexpected) companies is looking at how they can get involved to provide a solution and make a buck in the process. Let's look in more detail at what's been going on here.

Private and profit-oriented companies tend to head to where they think they have a solution for a problem, where they can devise a product for a new market, and where they can figure out a way to make a good return on their investment. Take for example Caterpillar, an industrial equipment manufacturer, and a company that on the surface doesn't seem to have much to do with water. One product that Caterpillar manufactures is a diesel-powered electrical generator. It turns out that many major cities in this country rely upon source waters that are lower in elevation than the city itself. In those cities, the water utility has to pump raw water into the treatment plants and distribution system. Those pumps are powered by electricity from the grid, but when the power goes out, the water still needs to flow. Several years ago during a power brownout in the Northeast, the city of Cleveland was essentially without water for several hours; a fire could have turned into a major catastrophe. That event was a red flag for a lot of municipal water providers. The leaders of those utilities quickly bought backup power generators from Caterpillar to be used in just such an emergency. Caterpillar is not what we'd think of as a water company, but it stands poised to provide equipment to the water industry, and with that derive sales and profits. The company also recently announced that for the first time ever it would offer two types of chassis for the "increasingly important market segment" of water trucks.

Caterpillar is also looking at its own internal water consumption and has determined that it uses more than 5 billion gallons of water a year in its fifty different facilities spread across twenty-three countries. It also figured out that 78 percent of those facilities have only a single source for the water needed to operate the plant. So now the company is looking at every one of those 5 billion gallons, trying to figure out what its own water risk is as a corporation and determine ways in which it might be able to conserve or reuse more water. (Water risk is an important concept we'll touch on later in the chapter.) To do that, the company will evaluate all of its vendors— the people and companies that Caterpillar does business with—to understand its bigger water risk picture and to help reduce its overall water footprint.

Or consider the case of one of Caterpillar's chief competitors, John Deere, maker of the iconic green and yellow tractors. Over the past few years, John Deere has embarked on a push to become a world leader in advanced irrigation systems, which lie at the heart of what will keep farms productive as water becomes scarcer and more valuable. Deere now has an entire division known as John Deere Water, which purchased a handful of smaller companies and now manufactures water-related irrigation equipment and supplies in ten countries and sells that equipment in more than a hundred countries. John Deere recognizes that the future of farming is directly tied to the future of water, and if it wants lots of farms around to sell tractors to, it is going to need to make sure that the farms have water and the right kind of irrigation equipment. This is just another example of a company that we don't really think of as a water company, but one that is gradually realizing the critical impact of water on its potential future business.

Hundreds, perhaps thousands, of smaller companies are also jumping into the water business, sensing similar opportunities. New entrepreneurial technologies and service providers are coming into the business from around the world, often emerging from regions where water problems are severe and where the needs are most immediate. For example, a number of notable new companies are developing in Australia. Also, many important developments have come, and will continue to come, from Israel, a tiny desert country where water is extremely scarce and valuable. Indeed, some leaders say that the major threat to Israel's security comes not so much from hostile Arab neighbors as it does from water scarcity.

Several smaller Israeli companies are poised to make a huge difference—not only at home, but also in the global marketplace. Take for example Netafim, which makes drip-irrigation equipment and other equipment for agricultural production, especially in Israel, but also in other countries. The company recently opened a new factory in Turkey. While the two countries do not share a religion, they do share a similar climate and the need to get as much agricultural production as possible from every drop of water.

These types of companies trying to make money in devising better products or systems to use water certainly won't be alone. A recent report from McKinsey estimates that the amount spent just on technologies to increase water efficiency will be $50 to $60 billion globally per year over the next twenty years. This spending will come from across industry, including agriculture, mining, consumer products, and more. This report, "The Business Opportunity in Water Conservation," sums up the situation well. "Making a business out of improving water efficiency won't be easy. Successful providers will have to migrate from selling equipment and components to selling solutions aimed at helping business customers reduce their water and energy use. The providers will therefore have to develop new skills and capabilities, particularly in marketing and sales, to identify and capture the higher-value-added solutions that business-to-business markets need. They must also engage more actively in shaping the regulations that will define this market—standing on the sidelines is no longer an option."[1]

The water situation is crying out for new technologies and approaches, and that's why we see so many new companies getting into the business. It's also why venture capitalists are now closely evaluating water deals that they probably wouldn't have looked at ten or fifteen years ago. There is another great indicator of the vitality and robustness of the industry: if you want to go to a conference about water, water savings, water treatment, investing in water, or about anything else related to the world of water, you won't have to wait long. Every week of the year, it seems, someone is hosting some type of water industry conference.

Other small water business opportunities abound for the creative entrepreneur, even though some of them may be fairly remote from the core water business. We found one fun example of that kind of opportunity being explored in Maine: the world has plenty of salt, and most Americans already get too much salt in their diets. But those who eat foods not processed in a factory need to add salt to their food. We've written plenty in this book about desalination, all of it related to getting the salt out of water so that the water is usable. One man in Maine is trying to get rid of the seawater so he can sell the salt. Stephen Cook started the Maine Sea Salt Company to sell salt from what he calls "the first salt works in Maine in two hundred years." He pipes filtered seawater into a series of greenhouses, then just lets the sun evaporate the water. After the water is gone, Cook packages up the salt, flavors some of it, and ships it all over the world to people who like the flavor, the potential health benefits, and just the idea of getting salt that hasn't been processed by one of the giant food-processing companies. Is this really a water business? It depends on your perspective, but it clearly got off the ground when an entrepreneur started thinking creatively about water and how to use it.

As we have seen earlier, the interplay of energy and water issues certainly represents one of the key areas of growth for the business in the future. Here again, many new companies are touting their abilities to save potential clients energy or water or both. Mineral and energy extraction companies will certainly be evaluating new products and approaches to save water. Power plants and agribusiness operators will gradually recognize that their water rights—their ownership of physical water—may be increasingly valuable, maybe even more valuable than the whole rest of their business. (We'll talk more about those markets later.) If, as we suspect, the type of solar power known as concentrated solar power gains market traction, we could see many more combined desalination and power plant arrays. (Concentrated solar power, which we wrote about in chapter 6, is the name for the array of mirrors that make minute adjustments through the day to pinpoint sunlight on a single point through which water flows and is instantly boiled.) In short, water and energy are increasingly just two pieces from the same puzzle, and in the future, we may see more and more companies claiming to be *watergy* companies.

WATER RISK

From a broader business perspective, virtually all companies are beginning to focus more on water issues and challenges. Even though they might not intend to participate in the water industry, most companies are increasingly realizing that water is critical in one way or another to the sustainability of their own businesses. This is especially true of those firms that need abundant clean water to manufacture their own products: industrial sectors such as food and beverage companies, semiconductor chip manufacturers, pharmaceutical companies, and others.

But the flip side of opportunity, water risk, also impacts many companies, even those who are remote from the water industry. Water is an input of some significance to almost all businesses, and the risks and challenges of growing scarcity are starting to appear on the horizons of almost all companies these days. According to a recent report from J.P. Morgan entitled *Watching Water*, the business opportunity and profit potential in water companies has been widely noted on Wall Street, but most investors have yet to really understand the business risks to companies that may occur in the future as a result of water shortages.[2]

Poor water management increasingly represents a serious risk to business success—and both corporate management and investors are beginning to take note of this. Companies are increasingly asking what would happen to their business if, for instance, water suddenly was rationed, or if its price went up five- or tenfold? For an individual family, that might mean giving up a nice green lawn. For a mining operation or a soft drink bottling plant, however, a disruption in the flow of water means that the operation might have to close down or move, possibly hundreds of miles away, to a water-rich location, with employees suddenly losing their jobs, and the company being forced to find alternative and higher-cost raw materials.

In reality, the situation is much more complicated than this for most companies. Few firms are yet effectively analyzing their risks up and down their supply chains. For example, a food processing company may have locked up the rights to the water necessary to operate a factory, but it may have no idea of the water challenges faced by its agricultural producers or other suppliers. If there are no potatoes, there's not going to be a potato chip factory, even if the factory has an excellent supply of water.

Even with the radical changes in irrigation systems and watering on farms that we wrote about in previous chapters, most freshwater used on earth in the future will still go to irrigate crops that are consumed either by people or by animals that we intend to consume later. And the business opportunities here are broad and diverse. Giant companies such as Monsanto are certainly researching water issues, and hope to develop seeds for crops that are either drought-tolerant or that can be irrigated with seawater. General Electric is now one of the world's largest manufacturers of membrane systems used to desalinate seawater and to clean and reuse industrial waste-waters. Although we may not think of firms like General Electric and Monsanto as water companies, the systems and expertise they develop and sell will have a huge impact on the future world of water. Many more companies, including some that nobody is looking at now, will continue to jump into this business in the future. In return for a financial profit, they will contribute some gadget, gimmick, idea, or service to the goal of solving world water problems.

WATER RIGHTS

We are also going to see new and radically different types of water investment opportunities in the future, new concepts about water that are only starting to emerge today. New concepts developed in the coming years will result in new opportunities and new companies. Let's take just one example that almost no one was thinking about just a few years ago: more and more private investors are getting interested in the concept of owning and trading actual, physical water, and we're beginning to see a new type of commodity market evolve for water. These markets, although still embryonic in most places today, are likely to become central to the overall future of the water business in many parts of the world. We should take a closer look at what's going on here.

Perhaps the best example of emerging water markets comes again from southeastern Australia, an area that has been suffering from an extremely severe drought for almost ten years. (Many experts forecast that the drought conditions experienced in Australia may be repeated in other parts of the world in the years to come; more on that in the next chapter.) Officials in the Murray-Darling Basin, the breadbasket of Australia, began to realize several years ago that there just simply wasn't going to be enough water to go around, and so they attempted to create a market trading mechanism whereby farmers and other users could actually buy and sell their water. To make a very long and complex story short, instead of simply paying the government a set fee for a fixed amount of water to grow their crops, users were given the right to a certain amount of water that they could use however they pleased. They could use that water to grow rice and alfalfa, or they could sell it to their neighbor. The government created a supply-and-demand-based system, almost like our over-the-counter stock exchange, where individuals can actually make those trades. Consequently, the price of water fluctuates day to day or week to week based on precipitation, agricultural commodity prices, and other factors.

With this trading mechanism in place, water quickly started to move toward those uses where it had the highest value: to the irrigation of crops that tended to allow the greatest profits to the farmers. The production of highly water-intensive rice, for example, quickly declined, with that water moving toward the production of higher-value grapes, pecans, and citrus fruits. While farmers may still wish they could get water more cheaply, the program known as Restoring the Balance seems to be achieving its goals of preserving the health of the water system and making the economy more sustainable even in the face of extreme and long-lived drought.[3] In short, the system made it easier to use scarce water where it had the most value and ensured that less water was wasted.

More and more policy analysts and political leaders are realizing that these types of mechanisms for trading water or water rights can be an important tool in improving water resource management and ensuring the efficient allocation of scarce water. Another interesting example involves the old Geneva Steel mill in Utah. This factory had produced steel since World War II and closed in 2001. The company that owned the assets of the mill sold off everything, including all the equipment, the tracts of land that encompassed what had been the mill, a mine, and other assets. All together, those assets brought the seller $101.8 million. Then the owner separately sold off the water rights associated with the physical landholdings. Those water rights sold for $102.5 million—more than the combined price of all the other assets.

Water ownership rules in the Western United States are based on the concept that the company or person that first put the water to use has the senior right to that water for all time. This system, widely utilized in the arid Western states, is called the doctrine of prior appropriation, or as it's often called, "first in time, first in right." The steel mill had secured the right to use the water decades before the population boom in Utah, and so it had the senior rights. Furthermore, subject to a very complex legal and transfer process, these rights can be bought and sold, oftentimes separately from the land that they might have originally been attached to.

Let's take another, hypothetical, example. Let's say there's a property in central Colorado near the Colorado River, a 200-acre ranch that's been active since 1910. Back then, the rancher staked a claim for 200 acre-feet of water each year out of the Colorado River, built some ditches, and flooded his fields to help grow the crops and grasses that nourished his cattle. Fast-forward a century, more or less, and the descendants of that rancher now want to subdivide that land to build some homes. They don't need as much water for that, so they decide to sell most of the water rights that have been attached to the property since their great-great-grandfather first starting ranching. The highest bidder for those hypothetical rights is a real estate developer 200 miles downstream in Grand Junction who is building a three-hundred-home property that needs a half an acre-foot per year for each of those homes for the next hundred years. That ranch family doesn't have the means to build a pipeline to ship that water to the thirsty developer downstream. But they don't have to. All they need to do is sign a piece of paper agreeing to sell the rights to the water, and let it flow on by their property so that the developer can instead pull the water out and use it far downstream. This kind of water rights trading is fairly common in the Western United States today, and it will probably become more common in other parts of the world in the future.

In general, most of the transactions going on in this area involve the transfer of water from agricultural irrigation to municipal and industrial use: drying up irrigated farms on the Great Plains and using that water to slake the thirst of the rapidly growing Western cities like Denver, Las Vegas, and Phoenix. And the price of that water is increasing sharply in most areas. Albuquerque is a good example of a trend that is being observed in many rapidly growing western cities, as illustrated in Figure 9-1 on page 168. But other interest groups, including environmentalists, understand how these laws work, too, and they are also getting involved in these markets for their own

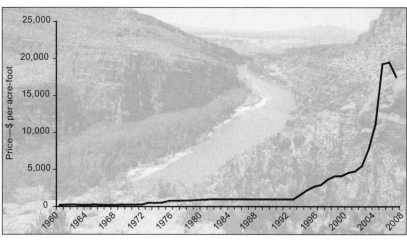

Source: Brown 2008.

FIGURE 9-1   NEW MEXICO'S MIDDLE RIO GRANDE WATER RIGHTS
PRICES, 1960 TO 2008

purposes. Groups such as Trout Unlimited and the Nature Conservancy are purchasing water rights from farmers or mining companies and then designating that they want that water to simply be left in the river—what are called *in-stream flows*. Water rights owners have always had to demonstrate that they are using the water for some "beneficial use," and for years the courts didn't see keeping the water in the river as a beneficial use. However, more recent cases have supported the concept of in-stream flows, and the courts are increasingly recognizing that preserving water in a river is just as beneficial as removing the water to use for a farm or factory.

States in the American West have been fighting over water since the day after they actually became states. Single court cases have stretched out over decades; it's not unheard of for lawyers to have spent the bulk of their entire careers on a single water case, as it bounced around among local, state, and federal courts and appeal processes. And the legal and political battles will continue. In a small drama played out during the 2008 elections, John McCain, the Republican nominee for president, offhandedly told a reporter that he thought we might someday have to reopen the Colorado River Compact, the agreement among the states along the Colorado River that determines which state gets how much water. All hell broke loose; Democrats quickly pounced on the statement, implying that it was a water grab by the Arizona senator to take water from other states. The event may not have gotten much attention in the East, but it certainly did in Colorado, Wyoming, Utah, and New Mexico—the upstream states. McCain quickly backpedaled, with the end result being that politicians from both parties in various states went on the record affirming their belief that the Colorado River Compact should remain as is. Of course, it's pretty clear that somewhere in the future we probably *will* have to reopen the compact, but the process will certainly be slow and arduous, and subject to huge political controversy.

In the Western United States (as in many other areas), water laws or policy frameworks that were designed and put in place more than a hundred years ago—and that may have made pretty good sense at the time—are starting to fail us today. Many observers believe, for example, that the prior-appropriation system leads to inefficiency and the wasting of scarce water in today's environment, that water rights holders are discouraged from conserving water in order not to risk losing the very right itself. Historically poor understanding of the connectedness between groundwater and surface waters also led to many laws and local policies that are clearly outdated today, yet remain on the books despite our improving knowledge of groundwater hydrology. Yes, the laws can adapt, evolve, and improve, but too many examples exist of archaic laws that tend to discourage the more innovative solutions or behavior that we clearly need to promote today.

The key interstate compacts themselves sometimes stand as giant obstacles in the way of more sound and innovative watershed planning, even though we now clearly understand that these agreements were conceived at a time when accurate measurements were not possible or, in the case of the Colorado compact, during an unusually wet period. The Western population, the economy, and the balance of political power are far different today than eighty or a hundred years ago, when most of these compacts were negotiated. But heavily entrenched and powerful interests have evolved in response to these systems as they were originally developed, and these parties will obviously and aggressively resist any kind of change in the status quo. Nonetheless, something will have to give eventually. We are eventually going to have to revisit some of these tenets of water law and policy, and it may not be pretty when it happens.

And it's not just in the West. Water shortages and water rights issues are increasingly a problem in the more humid Eastern United States as well, where the judicial system of riparian water rights was originally developed. *Riparian* means situated on the bank of a river, and the riparian system says that every landowner through whose land a natural surface waterway runs has the right to use that water as it flows by. An interesting battle recently shaped up in the Southeast when Georgia decided to try to redraw its boundary line with Tennessee, a line that had been established nearly two hundred years earlier. "This is a joke, right?" the governor of Tennessee said in 2008, when told of Georgia's plan. Georgia is claiming that the original boundary line was drawn incorrectly. And, surprise, surprise: the boundary Georgia wants to claim now moves things just enough to give it new access to the Tennessee River, potentially a new source of water, something that Georgia desperately needs. Rather than making serious efforts to reduce water usage in and around Atlanta, authorities instead are battling over the state line and the legal status of water from Lake Lanier, the traditional but rapidly shrinking water supply source for Atlanta. As one Georgia politician said, "I'm not going to lie to you, I want that water out of the Tennessee River."[4] This is probably another legal battle that will stretch out over years and years. In short, legal battles and water struggles are going to be increasingly common in the eastern half of the United States, mirroring what's been happening in the western half of the United States over the past hundred years.

Another proposed deal that may be an indicator of the kind of things we'll see more of in the future is the brainchild of legendary oilman T. Boone Pickens. While he's become quite famous for denouncing dependence on foreign oil and installing more wind farms in the United States, he's also started a water company. In the parched panhandle of west Texas, Pickens is hoping to take advantage of the laws in that state that allow unlimited pumping of water from any aquifer.[5]

In Texas, any landowner can pump as much water up as possible without regard for how it might deplete the aquifer or the problems it might cause for nearby neighbors. Often referred to as the *rule of capture,* this law is probably more properly described as "he who has the biggest pump, wins." (As of this writing, a case is in front of the Texas Supreme Court, *Edwards Aquifer Authority v. Day,* that may change that big-pump precedent, but the outcome is uncertain.)[6]

Basically, Pickens' plan is to pump water up from the Ogallala Aquifer, water that's been there for eons, and pump it a few hundred miles east to Dallas. Pickens created his own government utility district, an entity that has the power of eminent domain, so that he could build a pipeline to take all that water to suburban Dallas. So far the water utilities in Dallas have balked at the price of Pickens' water. As the price of water climbs, however, the people of Dallas will eventually have to pick between using fossil water from underneath the Texas Panhandle (or some other remote source), or cutting back on how much water they use. Pickens is banking on the assumption that people would rather pay a little more than actually try to cut back. Right now, Dallas uses more water per person than any other city in Texas.

Environmentalists will certainly make the case that Pickens' idea is a bad one, saying that the Ogallala can't withstand such a massive extraction. On the other hand, Pickens has other environmentalists singing his praises because of his commitment to wind power. Pickens has made some concessions on water, promising not to overtap in a way that might harm neighboring lands, but it's not clear that the plan will actually make much difference. One local water manager who opposes Pickens' plans told *Business Week* magazine, "It's like taking dollar bills out of your bank account and putting nickels back in. Even with a big bank account, there's an end. That's pretty much what's happening in the Ogallala."[7]

PRIVATIZATION AND
OUTSOURCING OF WATER

Finally, the whole issue of water marketing, water rights, and private water ownership leads us right into another topic that we should discuss in this chapter, a hugely controversial topic that sits right at the heart of the intersection between business and water. No discussion about the water business would be complete without taking a look at topic of *privatization*, and the related topics of outsourcing and the proper role of private capital in the delivery of public water. We wade into this discussion very carefully; highly emotional and contentious debates have raged around this topic for a long time and seem to be intensifying at the moment. Among the many divisive issues in the water market, none is as polarizing and controversial as the debate over the appropriate role of the private sector in the provision of drinking water. And, unfortunately, as with all disputes that tend to get emotional, the facts sometimes take quite a beating.

In one corner of the ring, we have those free marketeers and libertarians who think the water industry would be better off if local governments got out of it altogether, that the water business should be entirely privatized. They suggest that, in order to solve our water problems, we need to move toward viewing water as a commodity—just like wheat, copper, or oil—and allow natural market forces to set its price and more efficiently govern its allocation.

In the other corner, their opponents argue that water should be viewed as a basic human right, with the implication that clean water should somehow be equally and freely available to all. Unfortunately, this fight seems to be going the full fifteen rounds, and the angry participants usually generate more heat than light on the subject. In fairness, there is a good reason for this: these are complex issues that leave plenty of room for reasonable people to disagree.

In a little more detail, the basic arguments against privatization of water assets generally revolve around concerns about the perceived motives of private, for-profit firms, and the underlying belief that access to clean water should be equitably priced and provided to all. Most people understandably think that water resources are a critical part of our natural heritage. Some people therefore subscribe to the philosophy that public water supply and wastewater treatment systems should never be entrusted to private companies to own or manage. Some are concerned that private corporations will have little economic incentive to provide that kind of equitable access to water, and that they will act only in their very short-term interest.

Furthermore, people worry that once private operators have secured their contracts, they will overwhelm or control the municipal agencies that they contract to, or that private owners will somehow be able to influence the regulators who are supposed oversee and approve their operations. Finally, in some corners of the country, given the isolationist attitudes that have evolved in the United States since 9/11, the fact that many of these contract operators and water service firms are foreign owned tends to lead to even more concern and suspicion.

On the other hand, experience has shown that sometimes privatization or outsourcing may represent the best alternative for solving a complex water challenge. And the fact is, a number of fundamental economic drivers strongly support greater privatization in the water industry. For example, few municipalities enjoy overflowing coffers, and few public officials who wish to be reelected want to push through large tax increases on their watch. Most public works managers, particularly those in smaller municipalities, are between a rock and a hard place as technical requirements, regulatory complexities, and overall utility costs continue to climb. That's the rock. The general public remains very resistant to increasing taxes or users' fees to cover those costs. In some of those situations, the city agency may best be able to solve its challenges by turning to private industry.

It's important to realize that privatization is not a simple on-off situation; dozens of variations and gradations exist between true outright privatization and the simpler contract management or outsourcing of various parts of the process. Privatization may simply mean turning over some aspect of day-to-day operations to a private company under a long-term contract. In some cases, it may mean turning over the whole operation to a contract manager. In other cases, it can actually mean selling the water treatment and distribution infrastructure from the municipality to a private company, although in practice, this is quite rare.

In many cases, private companies *are* doing a better job at running the public water utility than a municipality might. Why is that? According to a person who runs one such private company, it's because the private enterprise system combined with public regulatory oversight does, in fact, work. Nick DeBenedictis is the CEO of Aqua America, one of only about ten publicly traded companies that operate water utilities. His business is focused around the suburban Philadelphia area. He points out that it's just good business, for example, to invest in water mains, on a planned and regular schedule. Far better to do that than to go in and make repairs after a catastrophic failure. Here's an analogy: it's like the difference between eating right and going to the doctor occasionally for checkups, versus eating poorly and then one day "suddenly" needing open-heart surgery.

Across the country and on average, various studies have shown that public water agencies are replacing less than 1 percent of their water mains each year, meaning they are on a schedule for replacement that stretches out more than a hundred years. On the other side, private companies are replacing approximately 1.5 to 2 percent per year. That means privately run water utilities are less likely to have catastrophic—and expensive—failures. In short, private utilities are reinvesting in the infrastructure at rates that are at least a bit closer to the real costs of sustainably delivering water, rather than postponing those expenditures for future generations. (We would argue, by the way, that given the neglect of recent decades, both public and private utilities should probably be rebuilding something more like 3 or 4 percent of their infrastructure each year.)

As should be clear by now, the public-private debate in water is full of these "on the one hand, but on the other hand" conundrums. But just as private operation may sometimes turn out to be the preferred solution, it's not as if all municipal systems are inefficient or in need of some kind of private assistance. On the contrary, it is clear that particularly the larger, fiscally sound cities are among the national leaders in terms of running efficient and profitable water and wastewater systems. It's often the smaller towns and villages that may need help from the private sector.

The desire to improve productivity and efficiency in municipal agencies was an early driver of the water privatization movement. However, after a decade or two of living under the potential threat of the city fathers bringing in a private firm, many public utilities have gone ahead and made substantial progress in productivity improvements and cost reductions, yielding the same results that privatization is touted to offer. One way or another, water and wastewater agencies are gradually becoming more competitive and efficient.

Before we complete this discussion, let's consider four important things about privatization, particularly given the way the topic is loosely tossed around in public debates. First, all public water providers *already* spend huge portions of their budgets on private contractors for crucial tasks, everything from initial testing of the water to running their public relations programs. So, it's not as if private participation in water treatment and delivery is some kind of new phenomenon. For example, when a public water utility installs a new water main, it will likely use software from a private vendor to help analyze the project bids from a range of other private vendors. It then purchases the necessary parts, equipment, and services from various private vendors. It locates the old main using a private vendor and then hires another private vendor to perform the main pipe installation or rehabilitation before the final private vendor comes to finish the job by paving the road on the top of the main. Yes, it's a public agency, but numerous private companies typically have a critical role to play in even the simplest expenditure.

Second, another important aspect of private operation that is also not widely recognized, particularly by the opponents of privatization, is that even where a private company completely takes over ownership of the water utility, it has essentially no control over water pricing and rate setting. In other words, in the case of a private operator, the company generally has no control of water pricing; that remains the responsibility of the city council or governing entity to which the private firm is simply a contractor. And in the case where a private company actually assumes ownership, it can raise prices only subject to strict oversight and regulatory control from the state public utility commission. Municipally owned systems can raise rates simply by internal board decision, but private utilities must go before the regulatory agency. The theory is that private companies must be regulated, but public agencies are in effect regulated by the public's ability to vote the managers out of office.

Third, more than 12 percent of the US population *already* receives its water from private organizations of one sort or another: publicly owned systems that are either *operated* by private contractors, or systems that are actually *owned* by private companies. Although there are hiccups everywhere, by and large most of these private systems have operated well for decades; in fact, the majority of their customers typically don't even realize that their water is being provided by a private company.

Fourth, in many parts of the world, private operation of drinking water systems is taken for granted, and in many countries it is the norm. In the United Kingdom, all of the water and wastewater utilities were consolidated together and privatized into about two dozen separate companies by Margaret Thatcher in 1989. Almost all of the water delivery in France has been done by private companies since the 1850s. It therefore seems ironic that the United States, which we like to think of as the home of free enterprise and the bastion of democratic capitalism, is so averse to private water systems while they are commonplace in countries that some think of as socialist. Indeed, it's more than just ironic: a recent European report accuses the United States of maintaining a "bunker mentality" in terms of being so close-minded to privatization—making impossible the solutions it could bring to a variety of water infrastructure problems.[8]

More than 45 percent of the population of Western Europe is now served by private operators.[9] But the trend is not just confined to Europe. Consider, for example, the city of Manila in the Philippines. When the government ran the water system, it had many of the problems associated with developing nations such as corruption and inefficiency. The authorities initiated a public-private partnership to manage the water utility, and now residents are receiving safe drinking water at an affordable price relative to the local economy. Compared to other similar-sized cities in other nations, the health of the people is much better than in those cities with government-run water utilities.

On the other hand, look at the publicly run Phnom Penh Water Supply Authority in Cambodia, which recently won an international award for the success it's had in dramatically lowering costs and getting safe drinking water for the first time to a huge percentage of the population, a significant portion of which had never seen water from a tap before. Obviously, a well-run public utility can succeed spectacularly.

This discussion reflects the fact that no simple, cut-and-dried answers are out there; there is no one-size-fits-all rule. Sometimes a private solution may work better, and sometimes a public solution may work better, and sometimes a combined solution may be optimal.

What's the common thread behind any successful water provider, public or private? The bottom-line answer is this: thoughtful, strategic leadership and tight operational management. These characteristics can exist in both private and public organizations, just as corrupt or lazy water managers can exist in a private corporation as easily as they can in a government bureaucracy. With effective management and leadership, it almost doesn't matter what the structure of the water entity is.

What we're more likely to see in the future is some kind of hybrid system: a water treatment and delivery system where both public agencies and private companies are involved. These combined approaches range all over the board, but are increasingly coming to be called public-private partnerships, or PPPs. One reason we may see more partnerships here is simple economic necessity. Bill Owens, the former Republican governor of Colorado who is now involved in water development, points out that it wasn't really politics that drove Europe to privatization; it just made more economic sense. Private water companies would have greater expertise and better access to capital. Owens believes that as water prices climb, privatization will become a more obvious choice for many of those fifty thousand medium-size and small utilities—again, particularly the smaller ones—and the privatization controversies of the past will gradually fade.[10]

It is hoped that in the future, leaders on both sides of this issue will stop arguing from ideologically extreme end points. As we've seen throughout this book, the problems are pressing and the time is short. Free-market advocates who reflexively disparage any and all government and municipal systems as inherently inferior or inefficient need to tone down their rhetoric and look for workable solutions. And so do the human-rights activists who say that all water quality problems or shortages are somehow the direct fault of private interference in the marketplace.

In the end, whatever our politics or inclinations, we have to be practical and remember that clean drinking water costs hundreds of billions a year worldwide to gather, store, treat, and distribute, and that we as a society somehow have to pay those bills. Despite what some activists may want to believe, clean water certainly isn't—and never will be—free. The fact that access to water is a fundamental human right doesn't mean that private capital and private companies shouldn't have a major role in providing water in the future.

So, all in all, the future looks very bright for the water business. Our predicament calls for bright and innovative ideas, new and advanced technologies, and more integrated approaches to solving the broader water challenge. As we've seen, a large and growing array of private companies is lining up to deliver these new systems and solutions, and to try to make a decent return on their investment in the process. The specific role of private business in addressing and solving water challenges will vary; some countries may rely more on governmental agencies while others will promote a greater role for the private sector. But in almost all cases, we are likely to see more private companies working hand in hand with public agencies, both to solve funding obstacles or operational challenges and to make sure that freshwater can be made available to all. The challenges are great, but so are the opportunities. Over the long term, the global water industry is poised to be one of the fastest growing and most sustainable businesses of all.

❧

[1] Giulio Boccaletti et al., "The Business Opportunity in Water Conservation." *McKinsey Quarterly*, December 2009. http://www.mckinseyquarterly.com/ The_business_opportunity_in_water_conservation_2483.

[2] JP Morgan Securities, Piet Klop, and Fred Wellington, *Watching Water: A Guide to Evaluating Corporate Risks in a Thirsty World.* World Resources Institute. March 31, 2008. http://www.wri.org/publication/watching-water.

[3] Australian Government: Department of Sustainability, Environment, Water, Population and Communities, "Restoring the Balance in the Murray-Darling Basin." http://www.environment.gov.au/water/policy-programs/entitlement-purchasing/index.html.

[4] Patrik Jonsson, "Drought-Stricken Georgia, Eyeing Tennessee River, Revives Old Border Feud." *Christian Science Monitor.* Feb. 15, 2008. http://www.csmonitor.com/USA/2008/0215/p02s02-usgn.html.

[5] Susan Berfield, "There Will Be Water: T. Boone Pickens Thinks Water Is the New Oil—And He's Betting $100 Million That He's Right." *Business Week* June 12, 2008. http://www.businessweek.com/magazine/content/08_25/b4089040017753.htm.

[6] Morgan Smith, "Lawsuit Could Determine Future of Groundwater." *The Texas Tribune.* April 22, 2010. http://www.texastribune.org/texas-environmental-news/water-supply/lawsuit-could-determine-future-of-groundwater/.

[7] Berfield, "There Will Be Water."

[8] See page xi of the *Pinsent Masons Water Yearbook 2006–2007*, Pinsent Masons, 2006, London. http://www.nedwater.eu/documents/PinsentMasonsWaterYearbook2006-2007.pdf.

[9] See for example page 28 of the SAM Study by Daniel Wild, Marc-Olivier Buffle, and Junwei Hafner-Cai, "Water—A Market of the Future." 2010. http://www.sam-group.com/downloads/studies/waterstudy_e.pdf.

[10] Interview with Bill Owens by Scott Yates, May 2010.

CHAPTER 10

THE FUTURE ROLE
OF WATER

It's sometimes easy to lose heart, looking at the dire water situation around the world today. Many observers believe that the simple fact is we already have way too many people on the planet. Water issues could be much more easily solved if we just had fewer people. And it is in fact increasingly reasonable to wonder if the pressures we are facing in terms of water resources will finally send us the unmistakable signal that we have reached the point of too many people on this planet. Will those challenges start to act as a control on population growth or even eliminate part of our existing population? In a sad but all too real way, the water situation is already a life-or-death issue in some places.

Let's take a concrete example. Bangladesh is a chronically poor country—one of the poorest in the world—and has been since well before it seceded from Pakistan in 1971. In the 1970s and 1980s, Bangladesh had some of the lowest life expectancies in the world. Children often didn't survive to see their tenth birthday, often because they had limited access to clean water. International aid and relief agencies poured into the country during these years to construct wells and install pumps to bring up clean water from underground aquifers so that villagers would not have to rely on badly contaminated surface waters. International aid agencies and the Bangladeshi government installed more than eight million wells across the country. In a massive publicity campaign to wean people away from their customary surface sources, the government encouraged people to use this newly available source of water. It appeared to work: deaths from dysentery and other waterborne illnesses fell dramatically.

What the aid agencies and the government didn't do, however, was test those new groundwater sources for naturally occurring poisons. It turns out that naturally occurring arsenic permeates the aquifers feeding about half of those eight million wells. Given the average concentrations of arsenic in these wells, it can take as long as twenty years for people to begin to feel and see the effects of drinking poisoned water. Once people know that they have arsenic poisoning, however, they learn that they are likely to die, that there is basically no cure. This is the unbelievable fate of hundreds of thousands, perhaps millions, of people in Bangladesh, not in some bygone historical period, but right now.

It's a tragedy on a scale so unimaginably large that it's hard to comprehend. *National Geographic* has called it possibly the largest mass poisoning in human history.[1] Now many of the same agencies that were involved with digging the original wells are back, trying to convince people not to use the tainted wells. Teams from the United Nations have painted 1.4 million wellheads red, and have tried to instruct people not to use them. This leaves much of the population facing the cruel choice of using arsenic-tainted well water that might kill them in ten or twenty years, or using contaminated surface water that could give them dysentery or kill their children within a few days. It doesn't seem like a choice any human being should have to face in the twenty-first century.

That's what we mean when we say that, on a global scale, water crises are already beginning to turn into controls on human population. Without safe drinking water, people simply die. The UN estimates that perhaps ten million people die every year from waterborne illnesses. The real number is probably a lot higher.

Although the Bangladesh situation is dire, it is unfortunately just one of many serious water problems that our planet faces. We've already talked in this book about a lot of the problems and challenges of future water resource availability. We don't want to be too much of a Chicken Little; plenty of books and stories out there can already scare the pants off of you. However, this is a book about the future, and the fact is we're facing a lot of water problems. And the breadth and magnitude of these problems will obviously dictate many of the future trends that we're likely to see in water supply and water usage.

Consider the fact, for instance, that many of the world's largest cities still dump the completely untreated sewage of millions of inhabitants directly into natural waterways or oceans. Indeed, one can only marvel at the natural treatment capacity of our oceans when we see how much raw sewage we continue to dump into them. Human sewage finding its way into open water channels in Dhaka was the underlying reason those aid workers in Bangladesh wanted to develop new groundwater wells. For those of us who live in places like the United States, a visit to sprawling coastal megacities such as Accra, Lagos, or Mumbai is absolutely shocking in this regard. Until you've walked the streets in these places, it's hard to imagine or really appreciate the scale of human sanitation challenges.

Two and a half billion people have no access to basic sanitation. This is a stunning figure: that's almost 40 percent of the world's total population.[2] The implications of this one basic fact are simply immense, not just in terms of human health, but also in terms of potential future social upheaval and political imbalances.

And water problems don't just exist in the Third World. We in the United States still seem to be willing to tolerate massive losses of clean drinking water as a result of our dilapidated infrastructure and decaying pipes. The USEPA says that 17 percent of the clean drinking water produced in the United States essentially vanishes out of our leaky pipes.[3] This is some of the best drinking water in the world; the same water could save millions of lives in poor countries, and yet in Boston one gallon in three seeps away unused. In London, as much as one gallon in two drains away out of those pipes under the city, some of which are still made of wood. The tab to fix all those pipes, as we discussed earlier, is somewhere between $600 billion and a trillion dollars, just for the United States, and the bill just keeps getting higher every year. One estimate we've seen for bringing the existing infrastructure around the whole world into shape is a staggering $41 trillion.[4]

That's what the water infrastructure picture looks like. What about the environmental side of the situation?

The USEPA estimates that more than 40 percent of American rivers are too polluted for swimming or fishing;[5] the UN has estimated that in China more than 75 percent of the rivers are heavily polluted. Dams aren't typically considered pollution, but nature did design rivers to flow, and dams stop rivers from flowing. Because people traditionally live near rivers, the reservoirs that form upstream from dams also force people to move. The best estimates hold that over the past seventy-five years, more than forty-five thousand large dams displaced some eighty million people. The vast recent dam constructions in China (which we wrote about in chapter 7) are only the most visible such projects.[6] The ecological and social impacts of these large dams can be long-lasting, while—as we now recognize in the case of major federal water projects in the American Southwest—their economically useful lives may be relatively short because of sediment infill, drought, evaporation, and various evolving regulatory requirements.

In other areas, our natural wetlands, ecologically designed by nature to regulate and clean our surface waterways, are increasingly being dried up by development and urban expansion. Also, many major aquifers around the world are being depleted, or mined, at a vastly higher rate than the rate at which natural processes can replenish them. The mining of the Ogallala Aquifer and other critical groundwater resources, as we talked about earlier, is potentially catastrophic for agriculture, the US economy, and the world food supply.

Many rapidly growing metropolitan areas, such as the Denver suburbs and the city of Albuquerque, have been exhausting irreplaceable deep aquifers for decades, and now find themselves struggling not only to better protect their existing source waters, but also to locate and acquire new source waters to ensure their futures. Some of these situations could turn into economic catastrophes in the relatively near term.

As a grim summary, it is now predicted that half of the world's population will live with chronic water shortages by the year 2050. In short, we are rapidly creating a situation of severe *water stress* in many parts of the world. Once again, it's hard to predict the future, but we *can* identify several broader trends that will have a significant impact upon future water challenges, including population, economic growth, energy, climate change, and general demographic trends. Let's examine each one of these.

POPULATION

First, and most critically, more people are likely to make the problem worse. If we have more and more people using that same amount of freshwater—an amount that is essentially fixed—at some point, something is going to have to give.

How many more people can we load onto the planet? That's a tough call, but most people who study population trends agree that the number of humans on the planet will continue to increase for the foreseeable future. The United Nations has three different estimates, one with the population continuing to grow, one with it gradually leveling off, and one, probably less likely, in which the population actually starts going down in about thirty years.

What's known for sure is that the human population on our planet has grown every year since the end of the Great Plague over six hundred years ago. The rate of growth has fluctuated, and some demographers think that it may have already peaked, with the highest growth rate coming in 1963. By some estimates, the world population could actually top out at about seven and a half billion people around 2040. Or the rate of growth could increase once again with the result that the planet could hit ten billion people by mid-century. Who knows? It will probably be somewhere in between those two estimates.

And what happens when we have eight or nine billion people? Will we have enough water to go around? We will certainly have enough water for everyone to drink, and probably enough for everyone to have something to eat. The problems start when we consider where people live versus where the water is. In the United States, water conditions are already tight in the Southwest, but cities like Milwaukee and Buffalo can handle plenty more people and a lot more factories without running out. Around the world, the story is much the same. Some areas are parched while others have abundant water. The problem is that the population and economic centers of today are not always where the water is, and large numbers of people can't just simply pick up and move.

On the other hand, the news is not all bad. A recent US Geological Survey study confirmed some positive trends that have been under way for several years now. Total water withdrawals—the total amount that we withdraw for use from both surface and underground waters—in the United States leveled off around 1975; given our still-growing population, this means that per capita consumption has actually been falling for about three decades now.[7] This *is* good news, and much of it is the result of more water-efficient agricultural practices and manufacturing approaches. And it's not just here: the French company GdF Suez recently reported that total water deliveries in France have been falling by about 1 percent per year for the past fifteen years.[8] But another reason for this trend is that we in the United States have effectively outsourced quite a bit of our water consumption. For instance, cotton for our growing population is much more likely to come from overseas now than it was in 1975, and the cotton that we used to grow here consumed plenty of water. The same goes for manufacturing; it takes water to make steel and water to generate the power to run the steel mill. If our steel is produced in China, that's a lot of water that isn't being consumed in the United States.

If the rest of the world can follow the United States in reducing consumption per person, then we will be better able to keep up with the changing water needs. But that's not going to be easy, especially for countries like China and India, where economic growth may still trump long-term environmental sustainability issues, and which have huge populations that are trying to move up and into the middle class. The question of consumption brings us to our second trend.

ECONOMIC GROWTH

As we've said, it's not simply the massive number of people that will make the water situation tougher; it is also the increasing standards of living and the increasing expectations of all these additional people that will make the water issue even more challenging. Why? Because almost anything a person can spend money on, with very few exceptions, takes water to produce. When you buy new clothes, the cotton or other fibers took thousands of gallons of water to produce. When you buy a new car, even an energy-efficient one, the carmaker used hundreds of thousands of gallons of clean water to manufacture the metal, plastic, rubber, and electronic components. Manufacturing a new smart phone requires water every step of the way, often highly purified water. Remember the two or three gallons of water that are used just to produce one gallon of ultrapure water that washes off the chip inside that phone? People climbing the economic ladder out of poverty in Bangkok, Baghdad, or Bangalore will all be increasing their water footprint as they make that climb.

Fortunately, that bad news is again coupled with good news. Many studies have shown that as families become more economically stable, they tend to have fewer children. So while a person climbing out of poverty may have a bigger water footprint, that person may also have fewer children than his or her parents did. In addition, as those people become better educated and more aware of their relationship to water and the environment, they may also learn about and implement ways of using less water and energy.

One of the trickiest areas, as we discussed in chapter 4, will be the interplay of improving standards of living, food consumption, and water usage. As people develop higher incomes, they tend to eat more meat. And as we saw, it takes a lot more water to produce meat than it does to produce fruits, grains, and vegetables. As water becomes scarcer, therefore, several things are likely to happen. First, the more water-intensive types of agricultural production will tend to move closer to natural water sources and areas where more rain falls. Second, meat producers will convert to animals that are less thirsty, like fish or chicken. Finally, with the help of genetic engineering, we will probably see new strains of both animals and plants that will be able to make do with less water. The bottom line here is that as standards of living increase, pressure on our water resources will rise, even if the population stays the same.

ENERGY

As a result of these first two trends—toward an increasing and more well-to-do population—energy consumption is also projected to rise sharply. Recent federal studies project an increase in energy consumption in this country of 40 percent by 2050.[9] Where will that new energy come from? Probably from a mix of existing fossil fuels, nuclear power, and various renewables: solar, wave, and wind. But a key constraint or cost consideration on every source of new energy— even solar—is water. Given the current rates at which water is consumed in the production of energy, it just doesn't seem like there's going to be enough to go around. And again, it works both ways: in many geographic areas, the flip side is that we won't have enough energy to keep the water flowing. See Figure 10-1 for a summary of the widely variable energy costs of treating different types of water to drinking standards. What's going to give?

Take, for example, the Central Arizona Project (CAP), which provides water to four out of five Arizona residents and to many farms in the state. The CAP is by far the largest user of electricity in

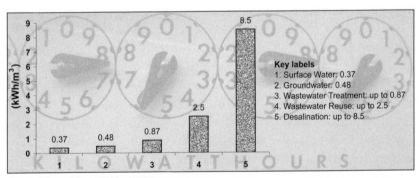

Source: Adapted from Webber 2008.

FIGURE 10-1 ENERGY REQUIRED TO PRODUCE CLEAN WATER (KWH/M^3) FROM DIFFERENT TYPES OF RAW WATER

the state: one huge coal-fired power plant provides most of the power to pump water from the Colorado River up nearly 3,000 feet in elevation. The USEPA is considering tightening the rules on smokestack emissions from that plant. If that happens, the plant will have to fund major new capital expenditures, and hence the local price of energy could jump dramatically. The money to pay for that energy comes directly from those who use water out of the project. The general manager of the Central Arizona Project recently told a reporter from the *Circle of Blue* newsletter that the power plant may have to close by 2019 if those rules are put in place.[10] This would obviously have a drastic effect upon Arizona's water supply. He may be trying to scare the government into backing away from these new regulations, or he may simply be stating the economic facts. This is just one example of how, at some point, for many energy-related companies, push will come to shove. Eventually, we will have to revisit some of our assumptions about life in the American Southwest. In other words, it may just not be possible, over the long term, to effectively deliver sufficient water to millions of people living in a high desert *and* keep the air over the Grand Canyon pristine at the same time.

If those energy costs jumped dramatically, it could essentially spell the end for large-scale agriculture in Arizona. Nearly every single farmer and rancher in the state benefits from the low-priced water they get through the Central Arizona Project or other diversion projects. Arizona farmers use that water to grow crops that thrive in the hot sun, but many of these are also particularly thirsty crops. The town of Yuma, for instance, is known as the "winter lettuce capital of the world." All that lettuce is grown with water diverted from the Colorado River. If the price of water jumps by a factor of ten, that capital will likely become a ghost town.

As wrenching as it is for the farmers and ranchers to hear, this may all have to come to an end. Beef may just have to come from Iowa, where water falls out of the sky for free. The amount of energy used to transport beef from Iowa to Arizona consumers is small compared to the amount of energy needed to transport the water uphill into a desert to water crops that can be fed to local cattle. Perhaps the question isn't really about why it is we should end agriculture in

Arizona. Instead, it may be: why did we try to develop agriculture here in the first place? Arizona is a desert state with an average of less than a half-inch of topsoil and no organic material in the ground. If agriculture in Arizona does decline, it will be because the combination of scarce water and high energy costs.

Let's take another, more direct case of the interplay between water and energy: the Marcellus Shale fields. This is the massive area of shale that sits under wide swaths of New York, Pennsylvania, and parts of nearby states. Trapped in that shale is enough natural gas to significantly alter the US energy independence equation. The problem? Water.

Drillers extract the gas using a technique called *fracking*, which is short for *hydraulic fracturing*. The hydraulic part, as you might imagine, involves mixing a lot of water with various chemicals and injecting it into underground wells under tremendous pressure. At least three distinct water problems emerge from this practice. First, fracking needs a lot of water. While this region certainly isn't under the kind of water stress that we find in the Southwest, no industry can suddenly start using thousands of acre-feet of water without having an impact on local water availability. Second, once the water is combined with chemicals and pumped underground, some experts think there's a danger of those chemicals finding their way into the groundwater aquifers and contaminating drinking water sources. Because some of these proposed fracking areas are not far from the New York City drinking water source area, they are getting especially close scrutiny. Third, a lot of the water injected into the wells, as well as some naturally occurring water, comes back up the hole in the gas-recovery process. That *produced* water is contaminated with those chemicals used in the process. The USEPA prohibits companies from dumping that produced water into the river—it has to be cleaned before being discharged or reused, or it must be transported long distances to be disposed of in deep injection wells. Once again, the Marcellus Shale may contain huge reserves of energy, but without solving the attendant water problems, it may never be recovered.

CLIMATE CHANGE

One of the biggest questions about future water availability, and perhaps an unanswerable question today, is just how global climate change may eventually affect the world water situation. We don't yet fully understand the breadth and severity of the impacts of potential global warming on the hydrologic cycle, but if global warming trends continue, we are likely to face a whole new range of challenging and potentially catastrophic problems.

The potential impacts of climate change on our existing water infrastructure could be dramatic. Global warming is expected to shift the *predictability, timing, and extent* of natural rainfall patterns around the world. This could truly wreak havoc in terms of water availability, agricultural productivity, and food supplies. The impact could be a sea change, in both the figurative and literal sense. While scientists are parsing the possible long-term effects of global warming, water policymakers and economists are just now starting to think about the potential economic and social effects.

Scientists seem to agree that in terms of water, the changes will include less precipitation in the areas that are already drier, like southern Africa or the Western United States. At the same time we'll see more precipitation in areas that already get plenty of rain during rainy seasons. That is, more rain will fall in areas that already have enough for growing crops—where it may create severe and unpredictable floods in densely populated areas. At least, they are populated now. That may change. As we've seen, New Orleans lost nearly half of its population after Hurricane Katrina.

Some of the localized changes predicted from global warming may look appealing in the short term; rivers fed by glaciers, for instance, may see higher flows in the near term for downstream residents, but that won't last long. If the glaciers melt altogether, the rivers could die.

Our entire national infrastructure for water storage and flood control, water treatment and distribution, and irrigation has been based upon historical climatic and meteorological observations. We've built reservoirs in certain areas and of certain sizes to provide sufficient storage to water local agriculture throughout the full year, and so on. If we begin to experience changes in the weather, all that may be either moot, or may not be enough to do the job. Dams and locks that work to hold back floodwaters may now be swamped in some cases or rendered meaningless in others. Reservoirs may overflow in the early spring but be dry by late summer, and so on. These may be extreme examples, but incipient climate change may just mean that our efforts to manage and live with water will need to be much more flexible and fast-moving in the future.

We can quibble about exactly what will happen, but this is scary stuff. The water resource managers who will look the silliest in the future are the ones who insist that the weather patterns of the past century are the best predictors of patterns over the next century. Flexibility and adaptability will be more important than ever.

GENERAL DEMOGRAPHIC TRENDS

Lastly, and perhaps most significantly, water considerations in the future may cause significant shifts in where people live, how they live, and how many of them survive—as we saw at the beginning of this chapter. All over the world, the trends are the same: big cities keep creating economic opportunities, and people migrate toward those opportunities. More and more people are concentrated in urban areas, which become dramatically short of adequate water and wastewater infrastructure. Fewer farmers work larger and larger plots of land.

But megacities unquestionably represent the future. Let's take another look at the most populous country in the world, with many of the biggest cities in the world: China. While China still has a large agricultural sector, all of the population migration patterns are toward the cities. New York City would be only the tenth largest city in China. The emerging Chinese cities are massive, and most of them have all of the water issues that go along with any big city. Infrastructure for drinking water and wastewater is a huge and growing concern that will likely become more acute as the cities and the economies grow. As we mentioned, rising out of poverty usually means an increase in per capita water use. In China, however, we won't see the benefit of family sizes decreasing with increased economic stability because of their long-standing history of limiting families to only one child. This means China is heading toward an era of increasingly severe water stress, even if the country does erect all the dams it currently has plans for.

And lest you think it's only the sprawling urban wastelands that have water and wastewater problems, take a look at the booming, ultramodern skyline of Dubai. Widely viewed as one of the most advanced and luxurious cities in the world, sporting the world's tallest building, largest shopping mall, and largest indoor ski resort, Dubai forgot to build one thing—an infrastructure system to treat its wastewater. The entire city relies primarily on a single wastewater treatment plant, 20 miles west in the desert. That plant is currently trying to handle twice the volume it was built for, and with almost no collection infrastructure to support it. All those fancy hotels and gleaming office buildings dump their wastewater into pipes that don't connect to anything; they simply end. So, all the waste collects in sump pits, and it then has to be pumped into trucks and hauled to the treatment plant. Take off from that sparkling Dubai airport, and you can look down and see a line of orange sewage tankers that stretches for miles. Drivers complain that they often have to wait as long as two days in that line to dump their sewage into the plant. And because they're paid by the load, some of them have decided to start dumping their sewage into gutters that drain directly to the sea, or just right on to the beach itself. This is not good for a city that is trying to attract the world's wealthiest people as inhabitants or as tourists to its beaches.

And that's just one of Dubai's problems; the emirate relies on energy-gulping desalination plants for practically all of its clean water. Hence it has one of world's largest carbon footprints, and on average, only four days worth of clean drinking water supply. By-product brines are dumped directly back into the ocean, and salinity levels have reached dangerous levels in coastal ecosystems. Although the emirate has transformed itself from a tiny desert port to one of the world's major cities in only a decade or two, its environmental problems are legion. The only good news here is that other booming cities in the region, like Abu Dhabi and Kuwait City, can learn from Dubai's mistakes.

In most countries, people have more freedom than ever before. That means people in general can do whatever they want to do and live where they want to live. Many people prefer to live in warm, sunny places, rather than in cold, overcast places, even if the cold places have plenty of water and the warm places are parched. That may not be quite so easy in the future. As water scarcity grows, we may see a long-term reversal in some of those historical migration patterns.

THE GENERAL OUTLOOK

So, what will these trends mean for the world of water in the future? Here are a few of our best guesses.

The one thing that's absolutely clear is that water prices will continue to increase. As we have more people using more water, we will eventually collide with the simple reality that the amount of naturally occurring freshwater in the world is fixed. It doesn't matter what your political philosophy is; increasing demand on a limited resource means the price will rise, one way or the other.

If you happen to live in a free-market or capitalist-oriented country, that means that you will be paying higher water rates, and you may also be paying more for the food you eat that contains a lot of water. In addition, you may be paying more for various types of taxes that go toward building and sustaining our water future. If you live in a country with a centrally planned economy, you'll be paying more, too; it may just be a higher percentage of your income going to government in the form of higher taxes. The bottom line is we'll all have to pay more for water, and that will induce us to try harder to use less of it. And not just in terms of direct water consumption but in terms of virtual water consumption as well—a much more difficult proposition. One way or the other, we're all going to pay more for water, likely dramatically more, and we're all going to have to use less. Water rates are already showing distinct increases worldwide, but as shown in Table 10-1, the price of water, as well as the per capita consumption of water, still varies considerably in different countries around the world.

TABLE 10-1   RISING WATER PRICES AROUND THE WORLD

Country	Average Water Price (US cents per gallon)	Per Capita Domestic Use (gallons per head per day)
Denmark	2.96	30.0
France	1.34	61.1
Germany	1.04	39.7
Australia	0.82	159.2
United Kingdom	0.69	36.6
Canada	0.64	204.7
Czech Republic	0.53	56.1
Turkey	0.53	62.6
Japan	0.48	98.2
Portugal	0.47	81.1
Spain	0.46	90.0
United States	0.43	162.1
Poland	0.39	39.2
Italy	0.31	127.1
South Korea	0.19	145.3
Mexico	0.19	52.6
Russia	0.16	96.8
China	0.11	25.0
India	0.05	36.6

Adapted from Global Water Intelligence 2010b.

Of course, water scarcity is not an evenly distributed problem around the world. Canada, for instance, has 7 percent of the world's freshwater and only 0.5 percent of the world's population. India, by comparison, has 4 or 5 percent of the world's water and 17 percent of the world's people. Saudi Arabia is more water-starved than India, but it has the money to deal with its water scarcity, at least for now. Some countries and regions have immediate crises on their hands; others may not face the same level of problem for decades or centuries.

Water will become a far more critical input or decision factor in all manufacturing and industry; it will increasingly be considered a factor of production in the same way that labor, capital, or energy cost inputs are today. We now happily pay the Chinese to produce many of the objects that we buy. A lot of global manufacturing has moved there, because labor and energy are relatively cheap, and because of China's lax environmental laws. But what if the Chinese government decides to stop damming its rivers and polluting them with industrial waste?

Water could become the limiting factor on Chinese industrial growth. It almost certainly will become a limiting factor in many other areas of the world at some point, in terms of future manufacturing, energy, and industrial production. For example, all that abundant natural gas underneath New York and Pennsylvania will stay right where it is unless we can figure out a way to extract it without using or polluting too much water.

Water may eventually be key in rewriting world history and reversing major social and demographic trends, both in the United States and around the world. We wrote in chapter 5 about the relative abundance of water in Milwaukee. The breweries and tanneries that helped that city grow in the early part of the 1900s are largely gone. Now city leaders hope to lure water-intensive businesses back to that city by building a sustainable water infrastructure. It may work, and as water becomes a more critical input, manufacturing may return to the city. Cities in what is now considered to be the Rust Belt may emerge as leaders in an economy that values water more highly.

Finally, water scarcity will increasingly become intertwined with national and international security concerns. We wrote earlier about armed water struggles in China in just the past few years; people were killed and plants were destroyed in skirmishes between regions within one country. We also wrote in chapter 9 about the struggle between Georgia and Tennessee; that one is being fought with lawyers, not bombs. Not yet, anyway.

Former UN Secretary-General Boutros Boutros-Ghali predicted back in the 1980s that the next war in the Middle East would be fought over water.[11] That particular prediction hasn't come true yet, but something of a cold war is brewing, not only in that region, but also in many other regions around the world. The journalist Steven Solomon points out the example of Yemen, recently a leader in producing terrorists to attack Americans around the world. As in so many other countries, people are leaving the barren countryside and moving into the big cities, mostly into that country's capital, Sanaa. Water riots are common there as more than half of the wells that produced water just a decade ago are now dry. The government tried to take away subsidies on the fuel used for illegal pumps, but after people took to the street to riot, the government backed down. Most of this illegally pumped water is used in wasteful flood irrigation to grow a plant that is used exclusively as a narcotic. Al-Qaida, Solomon writes, is able to recruit most effectively by simply digging wells.[12] That's an ominous connection between water scarcity and modern-day terrorism.

Regional water conflicts also have the potential not just to spawn terrorists but to disrupt the entire global economy. For example, manufacturing locations in India are becoming integral to the operations of hundreds of giant corporations, yet that country has very strained relations with Bangladesh and Pakistan, some of which are based on water conflicts. India and Pakistan are currently in the international Arbitration Court, arguing over a dam India wants to build high up in the mountains of disputed Kashmir, a dam that could potentially disrupt agriculture in Pakistan.[13] And as we mentioned earlier, even rumors of a possible dam in China that might block the flow of water into India could have those two countries quickly posturing for war.

Water issues are complicating or intensifying pressures in other trouble spots around the world: the Jordan River rises on the Syrian-Lebanese border and is used by Jordan, Israel, and the Palestinian territories. Water issues have contributed to shaping the current crisis in the Darfur region of Sudan. Nine often contentious countries depend upon the Nile River for most of their water. Any of these precarious situations could potentially lead to the kind of war that might drag in the United States or other countries. It may be that the best way to avoid the wars of the next hundred years is to aggressively improve the water situation of today.

But how can we improve the water situation enough to keep the peace on our planet? That's the big question, but certainly one part of the answer is to start to provide good information and better understanding of the issues. By reading this book, you've taken a step in that direction. In the next and concluding chapter, we'll get more specific in addressing some possible ways forward.

〜

[1] Pallava Bagla, "Arsenic-Laced Well Water Poisoning Bangladeshis." *National Geographic.* June 5, 2003. http://news.nationalgeographic.com/news/2003/06/0605_030605_arsenicwater.html.

[2] See World Bank, "40 Percent of World Population Do Not Use Improved Sanitation Facilities." July 19, 2010. http://data.worldbank.org/news/40prct-wrld-pop-dont-use-imprvd-sanitation, and United Nations, *Factsheet on Water and Sanitation.* International Decade for Action: Water for Life, 2005-2015. http://www.un.org/waterforlifedecade/factsheet.html.

[3] See for example Alliance for Water Efficiency, *Water Loss Control: Efficiency in the Water Utility Sector.* 2010. http://www.allianceforwaterefficiency.org/Water_Loss_Control_Introduction.aspx.

[4] Faisal Humayun, "Blue Gold: An Excellent Investment Opportunity." *Alrroya.com*, April 7, 2010. http://english.alrroya.com/content/blue-gold-excellent-investment-opportunity.

[5] USEPA, "President Clinton: Clean, Safe Water for All Americans." *Clean Water Action Plan.* EPA press release, February 19, 1998. http://www.epa.gov/history/topics/cwa/03.htm.

[6] See Jonathan Watts, "China at the Crossroads: Thirst of the Cities Drives the Giant Drills to Water China's Parched North." *The Guardian.* May 18, 2009. http://www.guardian.co.uk/world/2009/may/18/china-water-crisis.

[7] Andrea Hart, "U.S. Water Use Declines, But Points to Troubling Trends, Says USGS Report." *Circle of Blue* Oct. 29, 2009. http://www.circleofblue.org/waternews/2009/world/u-s-water-use-declined-from-2000-to-2005-part-of-growing-trend-usgs-report-finds/.

[8] Pinsent Masons, *Pinsent Masons Water Yearbook 2009–2010.* London, 2009. http://wateryearbook.pinsentmasons.com/PDF/Water%20Yearbook%202009-2010.pdf.

[9] Keith Schneider, "In Era of Climate Change and Water Scarcity, Meeting National Energy Demand Confronts Major Impediments." *Circle of Blue.* Sept. 22, 2010. http://www.circleofblue.org/waternews/2010/world/in-era-of-climate-change-and-water-scarcity-meeting-national-energy-demand-confronts-major-impediments/

[10] Brett Walton, "The Rising Cost of Settling the American Desert." Circle of Blue, Oct. 11, 2010. http://www.circleofblue.org/waternews/2010/world/power-plant-that-moves-torrent-of-water-uphill-considers-closing/.

[11] See James G. Workman, *Heart of Dryness: How the Last Bushmen Can Help Us Endure the Coming Age of Permanent Drought*, Walker & Co., 2009; and Patricia Kameri-Mbote, "Water, Conflict, and Cooperation: Lessons from the Nile River Basin," *Nile Media Network*, June 25, 2010. http://www.nilemedia-network.com/index.php?option=com_content&view=article&id=70:water-conflict-and-cooperation-lessons-from-the-nile-river-basin&catid=37:articles.

[12] Steven Solomon, *Water: The Epic Struggle for Wealth, Power, and Civilization.* HarperCollins, 2010.

[13] Ben Arnoldy, The Other Kashmir Problem: India and Pakistan Tussle Over Water. *Christian Science Monitor*, August 11, 2010. http://www.csmonitor.com/World/Asia-South-Central/2010/0811/The-other-Kashmir-problem-India-and-Pakistan-tussle-over-water.

CHAPTER 11

LOOKING FORWARD TO THE FUTURE OF WATER

Now that you've read this book, we are confident you'll agree that water will play a much more critical and central role in our lives and behavior in the future. Water will shape the full spectrum of economic, political, and social trends as well as our decision making. We will all begin to view water more as a precious resource and less as a free commodity to be exploited and wasted.

No matter how we might interact with and rely on water—as individuals, companies, people living within a watershed, or nations—we are destined to see sweeping and transformational changes in the not too distant future. We are hopeful that future perspectives on water will be more optimistic than our perspective today, and certainly more optimistic than the fictitious scenario presented in the prologue. If history teaches us anything, it is that future perspectives will undoubtedly be dramatically different, so maybe that's a good sign. But for things to improve, we need to much more aggressively embark upon solutions and actions now.

We hope the future of water *will* be one increasingly focused on solutions: new approaches, new technologies and systems, and wiser policies toward water use and conservation. We are optimistic that we can turn away from the head-in-the-sand, status quo approach that we have too often had up to this point. While the earlier parts

Solutions for Tomorrow

Four categories encompass the solutions available to us:
- *The importance of developing a broader and deeper public understanding of water issues*
- *Adherence to the philosophy of thinking globally while acting locally*
- *Pursuit of incremental technological advances and solutions*
- *Development of smarter laws and policies*

of this book have presented some interesting vignettes, concepts, and predictions about the future, let's face it: none of that is very useful unless we also discuss some solutions and ideas for improvement. We'll dedicate this last chapter to that effort.

Fortunately, the world water challenge *does* lend itself to small and incremental solutions and improvements in many areas. We can all begin to take steps to address water problems, even if some of them are baby steps. And we can also set in motion, both individually and collectively, many larger initiatives and solutions to address and reverse some of the bigger and more problematic trends and to begin to move in the right direction. Let's briefly examine a few of these recommendations and ideas.

Develop Broader and Deeper Public Understanding of Water Issues

Better public education programs about water are at the core of beginning to understand and proactively address the challenge. If people have a better understanding of the nature and seriousness of the problem—if they are better informed—then many of them will start to act more responsibly and make better decisions.

Understanding that recycled water can be clean and safe to drink, understanding the water implications of having a plate of beef versus having a plate of tofu, understanding the implications of flushing unused prescription drugs down the toilet, understanding that our infrastructure is crumbling, understanding that water prices may actually *need* to rise in order to provide safe and sustainable drinking water: the list could go on. As more people have a better understanding of the true scope of the water challenge and how they impact it, they will start to make better decisions.

Think Globally, Act Locally

Unlike most other resource problems and constraints, water challenges are often local or regional issues. While technologies, conceptual solutions, and broad policy approaches may be similar and applicable around the globe, implementation of those policies and approaches needs to be much more location specific.

If China burns cleaner coal, air pollution problems next door in Japan may lessen; however, better water conservation practices in Arizona aren't going to help to alleviate water shortages in southern India. Some areas don't really have any serious water problems, while others face a severe and immediate crisis. Although it's often been said that the Western United States has a water quantity problem, while the East has a water quality problem, we now recognize that the whole country has both quantity and quality challenges. More arid regions generally tend to face more serious and more immediate supply issues, even though (counterintuitively) it's not always reflected in current pricing. Water issues do vary widely from one watershed basin to another, based on levels of industrial activity, historical population patterns, and so on. Technological solutions such as desalination may be economically feasible in one area but not another, depending on alternative supply sources, local energy costs, or distance to the sea. Community-based systems and local implementation decisions will always work better than centrally dictated plans. One solution may work for us; a different one might work for our neighbors.

The local perspective is most likely to reflect the form of a single river basin or watershed. Within a river basin, all users tend to have a common problem or set of circumstances and, one hopes, a similar set of objectives in water resource usage and management. Within an individual watershed region, all users should be willing to address and pay for certain water quality standards and water availability. Water issues may be quite different or even nonexistent in a neighboring watershed basin.

It's worth noting that some 260 major river basins exist in the world, and these watersheds cross 145 national boundaries. Some 60 percent of the world's population lives within those 260 basins. As water becomes scarcer, these three simple facts point to the likely potential of serious political problems in the future.

Although we're probably not going to start redrawing national boundaries (our earlier example of Georgia's attempt to do just that notwithstanding), we should try to reform political and social institutions to reflect the more regional or local nature of water problems, establishing or streamlining systems that will work to coordinate national or international policies with regional decision-making bodies and localized solutions.

At the same time, we need to build international coalitions and frameworks and find new ways to cooperate, particularly from the virtual water perspective that we discussed earlier. An example of this is the recent purchase and development of agricultural capacity in wetter Central Asian countries by the parched but wealthy Arabian Peninsula countries—an acknowledgment that it makes more sense to grow crops where water is abundant. In short, while one solution may work here, a different one may work over there. And as we move forward, it is clear that some mix of top-down and bottom-up strategies must be devised for sharing, more efficiently allocating, and using water.

Pursue Incremental Technological Advances and Solutions

We've pointed out that there really isn't going to be a simple or major technological fix or silver bullet that will overnight and miraculously solve all our global water problems. However, incremental technological advance *is* ubiquitous. Technology *will* continue to march forward, and we *can* do much better in terms of applying, improving, and sharing technology and scientific understanding to better manage our scarce water resources.

Even without the iron hammer of higher prices forcing us to make more rational decisions, new technologies and systems can help us more efficiently produce and consume water. As we saw earlier, advances in soil moisture monitoring and smarter irrigation techniques are contributing important savings in agricultural water usage. More advanced and widespread residential metering technologies will help us all to be more careful and smarter about the way we use water at home. Techniques for in situ repair and extension of the life of existing infrastructure will mean less water loss and wastage and more water put to efficient use. Water catchment and harvesting, whether in the form of an individual tank on the roof of a house or new storm-water collection systems in San Diego, will prevent available freshwater from slipping away unused by humans. Storing water in underground aquifers instead of in surface reservoirs will reduce loss through evaporation, and the lining of earthen irrigation ditches will prevent water from seeping away.

In the first few chapters, we talked about technologies and systems that the individual can use, like low-flow toilets and showerheads, xeriscaping, or wiser lawn-watering practices. Other home conservation techniques have already considerably slowed down per capita water usage.

Desalination technologies could provide virtually limitless new sources of clean water, but energy and environmental questions and geographic limitations mean that this technology will only be practical in certain fairly restricted areas. Mobile seagoing desalination plants may be able to address some of these environmental questions and provide emergency supplies to population centers that are experiencing droughts or short-term breakdowns in infrastructure.

Clearly, vast opportunities exist to put new technologies, innovative management systems, and bright minds to work in addressing the world's water scarcity challenges. Advancing technology will be a critical component in starting to solve global water problems, but it must be exploited in combination with more careful and conservation-minded attitudes, greater efficiency of use, and smarter policies and management approaches.

But technology alone cannot be relied on to solve all our water problems. It may ease the pressure or delay the day of reckoning, but none of technology's bright solutions should lull us in to thinking that we don't need to work hard on water issues.

Develop Smarter Laws and Policies

Talk is cheap. It's easy to say that we need to reform the way in which we legislate and regulate water issues, and lots of people are saying that, but it's a whole different ball game to actually enact such changes and put them into practice.

It's apparent that in the United States, we need to refashion long-standing policies, regulations, and laws—indeed, our whole way of thinking about water. Government subsidies, major federally funded water projects, and interstate water distribution and irrigation programs over the past century were all undertaken for sound political or economic reasons at the time. However, they have also seriously distorted the workings of local market systems and have led to usage and allocation decisions that may not be in the best interests of the country now. Our current regulatory structure, while well intended, creates conflicts or unnecessary hardships and actually may not protect us against many of the contaminants and health risks it was designed to avoid.

We have just argued for the need to find specific, localized solutions to water challenges. However, many broad and overarching policy questions and legal issues must also be addressed in a more holistic manner at the national level. And many of these critical watershed-based concerns that don't follow political boundaries can only be solved by international cooperation. However, just to look at the case of the United States, the sheer number of different federal agencies involved one way or another with water tends to make policy coordination or change an almost impossible challenge. Presently, almost forty federal agencies or entities have authority over water issues, and they often act in isolation. These entities are spread across six cabinet departments, thirteen congressional committees, and some twenty-three subcommittees, all involved one way or another in water resource management. It's no wonder that federal water regulations and policies are sometimes confusing or contradictory, and all of this is before we get down to the level of regional, state, and local watershed authorities. Consolidation of these responsibilities into some sort of coordinated department or agency would help get rid of the so-called silos and make the job of holistically managing water resources easier: the concept of *one water* recently promoted by the Clean Water America Alliance.[1] Most politicians say such consolidation of power and control is unlikely, but at some point it's going to become critical.

The whole arena of water resource ownership—the legal framework of both the prior-appropriation and riparian water rights doctrines that we discussed in chapter 9—is coming more and more to the forefront of water resource use and management, particularly in the West. Here questions abound, and any type of change in the legal framework is staunchly opposed by agricultural and industrial users. But things may eventually have to change. For example:

- Does the prior-appropriation "first in time, first in right" doctrine still make sense today, or does it lead to inefficiency and the wasting of scarce water resources?
- Do the interstate compacts that were inked in the early part of the past century still make sense? People in the West simply can't imagine that these compacts could ever be reopened or changed, but someday they may have to be.
- How will these issues be affected by climate change? As we've discussed, this is one of the great unknowables of the water challenge.
- Is there any practicable political way to make significant changes?

Many observers have pointed out that we suffer not so much from an absolute shortage of water as from an inability to properly manage and allocate that water that we *do* have. This is a good perspective to keep in mind, but it requires us to resolve a lot of these intractable policy questions and to reformulate our political institutions. An authoritative report on global water issues from the *Economist* concisely concluded that water is "ill-governed and colossally underpriced."[2] Asit Biswas, a leading authority on international water management, has widely insisted that there is no water crisis, and that we have plenty of water to go around; we just need to get better at managing and using it.[3] But, as Peter Gleick of the Pacific Institute put it in a recent report, "There is a vast amount of water on the planet—but we are facing a crisis of running out of sustainably managed water."[4]

BACK TO OUR RECURRING THEMES

And now, finally, let's return to the four central concepts that we highlighted in chapter 1 and that we have tried to weave into the discussion throughout this book.

We Need to Balance Our Resource Trade-Offs

As we said at the outset, water is becoming a more critical issue for and key input to almost all personal and industrial decisions. But, despite our focus on water in this book, it's not the only factor or input that we have to consider in making economic or social decisions. Unfortunately, one seemingly logical and well-thought-out approach to more sustainable behavior may often be at odds with another.

We'll find that it's often not possible to minimize our carbon and water footprints at the same time. Buying asparagus grown in the Central Valley of California with scarce water transported from hundreds of miles away may not be very good for our water footprint. But buying asparagus grown in Peru and shipped by jetliner and truck to the local grocery store is not very good for our carbon footprint. What's a responsible consumer supposed to do?

Or, consider another example: the buy-local consumer trend that is emerging in many parts of the United States as a means of promoting local agriculture, eating healthier, and reducing the carbon footprint of large-scale food transportation around the world. The buy-local movement, while it has many attractive aspects, may often be in direct conflict with the concept of water footprint or the local availability of actual water. Does it really make sense to use up very scarce water trying to grow vegetables in the desert outside of Santa Fe so that wealthy residents can enjoy the satisfaction of buying food-stuffs at the local farmer's market? If one looks around at many of the major and growing cities in the Southwest and around the world, there simply isn't sufficient water or the appropriate climate to grow locally all the needed food.

And it's not just water or energy considerations that go into these difficult decisions and trade-offs. The other inputs or decision factors that we have been highlighting also enter into the equation. Labor costs and labor conditions are often issues. The capital costs of making something can often vary according to location because of widely variable environmental regulations; that's largely why we've seen so much mining and manufacturing move out of the United States. Geopolitical, moral, and ethical considerations can also cloud and complicate these types of decisions. Should we buy running shoes made in a plant in Asia with poor working conditions when boycotting those shoes may put the plant out of business and drive those employees into even deeper poverty?

Sometimes, trying to carefully evaluate a decision or a behavior from the perspective of all of these critical inputs can lead to some very interesting, counterintuitive, or even humorous conclusions. When attempting to take all of these factors—energy consumption, food consumption, implied water and carbon footprints, and so on—into account in carrying out routine daily tasks, some researchers have come to some rather surprising findings. For example, it's been suggested that in some cases it may be more environmentally sustainable to drive your car to the store to pick up a few items than it is to ride your bicycle or walk!

How can that be? Let's say you live in Norway, close to abundant fossil fuel production. Most of your food has to be grown far, far away, say on farms in Spain that have to be irrigated and treated with chemical fertilizers. Those water and energy-intensive foodstuffs are then flown in high-carbon-footprint jets to Norway to provide your body with enough energy to walk or ride your bike to the store. Taking all of these various concerns and inputs into consideration, researchers can show that it's better to just hop in the car if you need something from the store and save all the energy needed to ride your bike there. And, as you might guess by now, this effect is even more pronounced depending on whether you're a vegetarian or if you get your sustenance more from eating beef.

And of course, if you don't ride your bike to the store, you won't have to use so much water to wash all your sweaty clothes, and you won't have to dump as much phosphorus into the sewer from your detergents. But then again, water is plentiful in Norway. You can begin to see the complexity of really looking at an issue or a given behavior from a broad environmental-sustainability perspective. What is a unit of water worth versus a unit of energy or an avoided bit of carbon into the atmosphere?

In summary, one idea, approach, or philosophy may appear very logical or elegant when viewed in isolation, but when it is viewed from a more holistic and integrated perspective, it becomes clear that many different approaches and objectives have to be considered and balanced. As we step back and take a more global view, we begin to understand that everything is tied together. We can't view some of these issues in isolation. For any individual person in any specific place around the globe, our carbon footprints, our water footprints, our agricultural footprints, and food consumption are all tied together in different, complex, and intriguing ways.

We Need to Better Incorporate Virtual Water Thinking into Our Behavior

Despite the foregoing paragraphs, this book *is* about water, and from that perspective, we must start to incorporate the concept of virtual water into more of our trade, consumption, and commerce patterns. An international system to promote export of more water-intensive foodstuffs such as rice to relatively drier countries can free up water there for other more critical uses, and perhaps create a more stable political situation in the process. The liberalization of agricultural trade policies and tariffs is obviously a vexing political challenge, but progress here could contribute to better production decisions and ultimately to the individual competitive advantage of nations.

At the same time, the concept of virtual water has serious limitations and may in some cases conflict with other trade or consumer objectives. Because food requires so much water, international trade patterns in virtual water are essentially a reflection of trade patterns in agricultural commodities. Stronger industrial countries without as much agriculture will obviously tend to be net importers of water in the form of food, whereas less industrialized and more agrarian countries will tend to be agricultural (and water) exporters, regardless of their natural water resources. Nonetheless, as Christopher Gasson of *Global Water Intelligence* put it, "You cannot tell peasant farmers in North Africa or India that they should give up their land and become advertising executives or bank clerks because those professions use the least water."[5] Perverse virtual water flows are here to stay, and what really needs to be addressed, as we have emphasized throughout, is the efficiency of that water use where it is most scarce.

As we noted earlier, better conservation practices in Arizona are not going to help solve water problems in southern India. However, changing purchasing habits in Arizona might help solve water problems in southern India. Shifting food consumption behavior could have a major impact on water availability in specific regions of the country. These are big issues, and things are not likely to change overnight, but a better understanding of our real water usage will allow us to at least consider making better decisions.

We Need to Think More Holistically about Water

We have talked about lots of different types of water throughout this book—drinking water, wastewater, storm water, and so on—but we hope we've also demonstrated that, in the final analysis and from a more holistic perspective, all of these different types of water are all just simply one thing: water.

Too often, however, we all still think and behave as though water were defined and characterized by all these different labels. Too many of us still think of ourselves as "storm-water managers" or "drinking-water guys" or "wastewater experts." Storm waters and sewage are still thought of as a problem or wastes to be disposed of, not as potential resources to be harvested and productively used. Groundwater users are still treated to a different set of legal and regulatory requirements than surface-water users, even though we now understand that surface waters and groundwaters are often interconnected. And these perspectives and problems are unfortunately reinforced by an increasingly archaic and often conflicting set of federal and state laws, a plethora of congressional and legislative committees with disparate jurisdictions, and numerous federal and state water agencies each of which has just a single purpose or mandate. And it's generally the same situation around the rest of the world.

We now understand that not only are most of our water problems interconnected but that they are also interrelated with many critical issues beyond water: energy supply, air pollution, urban development, endangered species, transportation and housing, and so on. The more we learn about a given water problem, the more often it requires us to stretch our thinking outside the traditional mind-set of water sector professionals. We need to move beyond this patchwork type of approach. More thoroughgoing and comprehensive reforms are required. Our water policy is now too critical to be defined or governed by these types of historical exemptions, exceptions, and additions.

The Clean Water America Alliance has recently worked to more broadly publicize this concept of *one water*, and to underline the fact that this type of historically constrained thinking is the major cause of dysfunction as we try to formulate more of a national water policy.[6] The proverbial silos or stovepipes of different and often conflicting stakeholders may have made some sense at one time in the past, but in a collective sense they are now woefully outdated. And while we may be starting to grasp the concept of one water, we still don't usually act and behave in that way. Of course it's much easier said than done, but we need to get rid of all these little tails that are still trying to wag the big dog. We need to think outside of these silos about all of our different types of water and begin to consider them all as just one water.

Think for a moment about an astronaut circling in a spaceship far above, gazing out her window and down at our spherical little planet earth. From that perspective, it's pretty clear that we are separate and self-contained little ball, mostly covered with water: a closed system, a zero-sum game, an isolated and solar-powered desalination plant quietly floating through space. We need to think of our water resources like that astronaut might. We have a lot of water: most of it is in the ocean; some of it is raining down over the continents in various places; some of it is flowing down rivers and streams; some of it is sitting quietly in underground aquifers or polar icecaps; some of it is dirty and waiting to be cleaned up; and some of it is flowing through our houses, businesses, and bodies at the moment. Each one of us uses some of those molecules of water; we will make some of it dirty, we will clean it up again, and someone else will use it later. We can't create new water, and we can't destroy it; it's all just there: water.

We Need to Prepare for the Inevitability of Rising Prices

And finally we return to the key concept underlying this book: the need for more realistic pricing of water. We don't like to repeat ourselves, but let's say it one more time. Water prices simply must rise, not only to better reflect true cost and value, but also to help facilitate many of the necessary changes in thinking, policies, and usage that need to occur. If one single and inescapable conclusion results from any review and discussion of the world water situation, it must surely be the inevitability of continuously rising water prices over the longer term, indeed, the *urgent need* for rapidly rising water prices in many parts of the globe. As that happens, so many of the other changes that we have discussed in this book will follow.

Water has traditionally been priced so low that most users simply don't have any serious economic incentive to conserve it or use it wisely. People naturally don't pay much attention to or conserve a commodity if they tend to view it as virtually free—and until recently, that is exactly the way in which a lot of people viewed water. And too many politicians around the world tell themselves, "If you want to stay in office, then provide people with free water."

The true cost of delivering clean water, as well as the average price of water, *is* continuing to creep slowly upward in most localities, but in most areas, governments have not allowed prices to rise to the kinds of rates that will be necessary if we are going to upgrade and maintain our infrastructure on a truly sustainable basis. Almost all water utilization decisions and resource management issues would be far more efficient, and solutions would begin to emerge more quickly, if water prices were higher.

As prices rise, decisions about water usage will inevitably begin to take on greater significance in the overall economy, and many of the incipient trends we have discussed will gather steam: greater reliance on reuse and recovery, more emphasis on conservation, a continuing trend toward more private-public partnerships, and more rapid advances in technology.

Among observers and water policy leaders, there really isn't a lot of dispute about this. The key policy question here is not really *whether* prices should rise, but more *how* they should rise: gradually and according to the market forces of supply and demand, weak though they are, or through some sort of government mandates and policies.

However, the critical flip side of this coin is that higher water prices also inevitably raise the issue of the ability to pay by different people all across society, and the questions of whether and how subsidies should be provided to certain parts of the population. This is an issue that may not be adequately addressed by market mechanisms and that must receive careful attention from federal and local policymakers. Indeed, one of the great challenges of the future of water will be trying to simultaneously treat and manage water more like a commodity, while at the same time recognizing that access to water is a fundamental human right. In the United States, we've tried to work out that challenge with regard to food through the use of food stamps and federal and state programs, and we'll probably need something similar in terms of ensuring an adequate access to water for all.

We still don't really recognize the true value of water, and few of us currently have to pay anywhere near what that water is really worth to us. Indeed, to quote that (overused) dictum of Benjamin Franklin from more than two hundred years ago—"When the well's dry, we know the worth of water."

CONCLUDING THOUGHTS

Imagine for a moment that an international commission of astronomers discovers that a massive asteroid is hurtling directly toward us and is certain to destroy the earth in ten years. With an immediate and coordinated international effort, however, the scientists say we have a decent chance to develop the technology to redirect or explode the asteroid before it destroys us. Because doing nothing means sure annihilation, the peoples of the world quickly drop their religious and political quarrels and agree to throw all of their resources and energy together to find a way for the human race to survive.

We need to develop this kind of mind-set with respect to the impending world water crisis. Other than air, there simply is no substance more critical to life than water; we cannot live without it for more than a few days. Processed clean water has made possible our advanced industrial economy and increased standards of living for the world's people. Modern irrigation techniques have allowed us to feed our expanding population and to turn deserts into fields of green. Yet we continue to deplete and pollute our limited water resources at an alarming rate, and we steadfastly look the other way while our water treatment and distribution infrastructure continues to crumble.

The twin challenges of water quantity and water quality represent an inexorable planetary crisis, perhaps the defining crisis of the twenty-first century. This may not have the sudden impact of the asteroid, but its ultimate effect may be just as dire. And fragments of the asteroid are already hitting us: people are already dying every day as a result of water scarcity and quality problems.

Yes, water frequently falls from the sky. Yes, three-quarters of our planet is covered with water. And yes, freshwater is abundant in many parts of the globe. But it's not always clean, it's not always where we need it, it's not always there when we need it, and it costs hundreds of billions of dollars a year to collect, clean, and distribute. The world's population has increased fourfold over the past hundred years, but we still have the same amount of water. And, unlike any other commodity, there is truly no substitute for water.

Serious water problems *are* hurtling toward us in the near future, and we need to take more dramatic steps right now to begin addressing these problems. Luckily, we don't have to wait until we are facing catastrophe on all sides; we can start to do something now.

છ

[1] See various reports from the Clean Water America Alliance, including *Managing One Water: The Clean Water America Alliance's National Dialogue Report*. http://www.cleanwateramericaalliance.org/pdfs/managing%20one%20water.pdf.

[2] John Peet, "A Survey of Water: Priceless." *Economist*, July 17, 2003. http://www.economist.com/node/1906846.

[3] See speeches of Asit Biswas at the 2009 Nobel Conference of Gustavus Adolphus College, http://gustavus.edu/events/nobelconference/2009/biswas-lecture.php, or *Transforming the World of Water: The Book of the Global Water Summit 2010*, Paris. Global Water Intelligence. http://files.globalwaterintel.com/dl/20101018_Transforming_the_world_of_water.pdf.

[4] Pacific Institute, "Peter Gleick Discusses 'Peak Water,' China's Water Crisis, Climate Change Impacts." Press Release, Jan. 5, 2009. http://www.pacinst.org/press_center/press_releases/woodrow_wilson_2008_2009.html.

[5] See Christopher Gasson, "A New Virtual Reality." *Global Water Intelligence*, 9:9 September 2008, http://www.globalwaterintel.com/archive/9/9/analysis/a-new-virtual-reality.html.

[6] See Clean Water America Alliance at http://www.cleanwateramericaalliance.org.

Acknowledgments

Thanks go first to my collaborator, Scott Yates, who did a lot of the legwork and background research, particularly for the middle sections of this book. Scott's energy and enthusiasm for the project helped us to get things moving, to broaden and round out the content, and to produce a more lively and readable book. I also thank the various people who worked with me from the American Water Works Association (AWWA): first, Scott Millard, who pushed me to get started on this project a couple of years ago and coordinated the project within AWWA; Martha Ripley Gray, who edited the manuscript; as well as John Anderson, Cheryl Armstrong, John Kayser, Roy Martinez, Deirdre Mueller, Dave Plank, Gay Porter DeNileon, Robert Rosamond, and Polly Wirtz. An additional thank-you goes to AWWA Executive Director David LaFrance, who took a special interest in this project and suggested a number of good ideas and additions. Thanks also to reviewers Bernie R. Bullert, manager of water/wastewater at TKDA; Thomas J. Lane, vice president, Malcolm Pirnie Inc., New York; Shane Lippert, senior project manager, Archer Western Contractors; and Ed Means, Malcolm Pirnie Inc., California.

A number of my professional colleagues and business associates were kind enough to take the time to read parts or all of the manuscript and give me helpful comments, including, in alphabetical order: Bruce Babbitt, former US secretary of the Department of the Interior and former governor of the state of Arizona; Bill Bertera, former executive director of the Water Environment Federation; Dick Champion, director of the Independence, Mo., Water Pollution Control Department and chairman of the Clean Water America Alliance; Ben Grumbles, former assistant administrator for water of the US Environmental Protection Agency and current executive director of the Clean Water America Alliance; Jack Hoffbuhr, former executive director of the American Water Works Association; Bill Owens, former governor of the state of Colorado; Jeremy Pelczer, chairman of WaterAid International and former CEO of Thames Water

Company; and George Stubbs, editor of the *Environmental Business Journal*. From among my colleagues at Summit Global Management, I would like to thank Rob Steiner and Tim Walsh for their critical review of the manuscript, as well as my friend and colleague John Dickerson for his ideas, suggestions, and encouragement over the past decade. A number of friends and colleagues who are interested in water issues and challenges also provided useful input and comments, including Steve Cox, Richard Harvey, and Jim Shaw.

A special thanks to Joyce Maxwell, who helped me with a detailed first-cut editing of the manuscript. Finally, thanks to my wife, Susan, for her continuous support and for cheerfully putting up with my frustrations and complaining, occasional long hours, and constant travels around the water business.

—Steve Maxwell

A Denver writer, Gene Fowler, working nearly a hundred years before me, said: "Writing is easy. All you do is stare at a blank sheet of paper until drops of blood form on your forehead." Other than the "paper," not much has changed. Still, I'm so glad to have had the chance to be a part of creating this book and have nothing but thanks for AWWA and Steve Maxwell. Also, to my family and friends who endured a year of me muttering about water, writing, and word counts... Thanks.

—Scott Yates

BIBLIOGRAPHY

Abbey, Edward. *The Monkey Wrench Gang.* Lippincott, 1975.

Alliance for Water Efficiency. 2010. *Water Loss Control: Efficiency in the Water Utility Sector.* http://www.allianceforwaterefficiency.org/Water_Loss_Control_Introduction.aspx.

American Prospect. The Global Freshwater Crisis and the Quest for Solutions. Special Report: June 2008. http://www.prospect.org/cs/archive/view_report?reportId=68.

Annin, Peter. *The Great Lakes Water Wars.* Island Press, 2006.

Arnoldy, Ben. 2010. "The Other Kashmir Problem: India and Pakistan Tussle Over Water." *Christian Science Monitor,* August 11. http://www.csmonitor.com/World/Asia-South-Central/2010/0811/The-other-Kashmir-problem-India-and-Pakistan-tussle-over-water.

Aspen Institute. Sustainable Water Systems: Step One—Redefining the Nation's Infrastructure Challenge. http://www.aspeninstitute.org/publications/sustainable-water-systems-step-one-redefining-nations-infrastructure-challenge.

Associated Press. Drugs in the Drinking Water. An AP Investigation: Pharmaceuticals Found in Drinking Water. http://hosted.ap.org/specials/interactives/pharmawater_site/.

Australian Government, Department of Sustainability, Environment, Water, Population and Communities. Restoring the Balance in the Murray-Darling Basin. http://www.environment.gov.au/water/policy-programs/entitlement-purchasing/index.html.

Bagla, Pallava. Arsenic-Laced Well Water Poisoning Bangladeshis. *National Geographic.* June 5, 2003. http://news.nationalgeographic.com/news/2003/06/0605_030605_arsenicwater.html.

Barber, Dan. How I Fell in Love With a Fish. TED Blog. March 10, 2010. http://blog.ted.com/2010/03/10/how_i_fell_in_l/.

Barlow, Maude. *Blue Covenant: The Global Water Crisis and the Coming Battle for the Right to Water.* McClelland & Stewart, 2007.

Barringer, Felicity. 2010. Lake Mead Hits Record Low Level. *New York Times,* October 18. http://green.blogs.nytimes.com/2010/10/18/lake-mead-hits-record-low-level/.

Barringer, Felicity. A Greener Way to Cut the Grass Runs Afoul of a Powerful Lobby. *The New York Times,* April 24, 2006, sec. National. http://www.nytimes.com/2006/04/24/us/24lawn.html?_r=1.

Bastasch, Rick. *The Oregon Water Handbook: A Guide to Water and Water Management.* Oregon State University Press, 2006.

Beckett, J.L., and J.W. Oltjen. Estimation of the Water Requirement for Beef Production in the United States –71 (4): 818. *Journal of Animal Science.* http://jas.fass.org/cgi/content/abstract/71/4/818.

Bell, Thomas J., and International Deep Waterways Association. *History of the Water Supply of the World: Arranged in a Comprehensive Form from Eminent Authorities, Containing a Description of the Various Methods of Water Supply, Pollution and Purification of Waters, and Sanitary Effects, with Analyses of Potable Waters, also Geology and Water Strata of Hamilton County, Ohio, Statistics of the Ohio River, Proposed Water Supply of Cincinnati* ... P. G. Thomson, 1882.

Berfield, Susan. There Will Be Water: T. Boone Pickens Thinks Water Is the New Oil—And He's Betting $100 Million That He's Right. *Business Week,* June 12, 2008. http://www.businessweek.com/magazine/content/08_25/b4089040017753.htm.

Bird, M. 2003. Supplying the Southwest with Seawater Purification. *Water Conditioning and Purification International.* (Dec): 40-49.

Biswas, Asit. 2009 Nobel Conference of Gustavus Adolphus College. http://gustavus.edu/events/nobelconference/2009/biswas-lecture.php.

Boccaletti, Giulio, Merle Grobbel, and Martin R. Stuchtey. The Business Opportunity in Water Conservation, *McKinsey Quarterly,* December 2009. http://www.mckinseyquarterly.com/The_business_opportunity_in_water_conservation_2483.

Bormann, F. Herbert, Diana Balmori, and Gordon T. Geballe. *Redesigning the American Lawn: A Search for Environmental Harmony.* Yale University Press, 2001.

Borrell, Brendan. Sausage without the Squeal: Growing Meat Inside a Test Tube. *Scientific American.* March 31, 2009. http://www.scientificamerican.com/article.cfm?id=test-tube-pork.

Boyle, Rebecca. Better Tomatoes Via a Fertilizer of ... Human Urine? Popular Science. http://www.popsci.com/environment/article/2009-09/fertilizer-future-might-be-closer-we-think.

Brady, Todd. A Water Policy? March 17, 2010. *CSR@Intel: Putting Social Responsibility on the Agenda.* http://blogs.intel.com/csr/2010/03/a_water_policy.php.

Braunstein, Glenn. Prescription Tap Water: What Drugs Are We Taking with Our Drugs? *Huffington Post,* Jan. 19, 2011. http://www.huffingtonpost.com/glenn-d-braunstein-md/prescription-tap-water-wh_b_809870.html.

Brean, Henry. Mulroy advice for Obama: Tap Mississippi Floodwaters. *Las Vegas Review Journal.* Jan. 12, 2009. http://www.lvrj.com/news/37431714.html.

Brown, F.L. 2008. The Evolution of Markets for Water Rights and Bulk Water. 53rd Annual Water Conf. Proc., New Mexico Water Resources Research Institute, Albuquerque, N.M.

California Energy Commission, California's Water-Energy Relationship, November 2005. http://www.energy.ca.gov/2005publications/CEC-700-2005-011/CEC-700-2005-011-SF.PDF.

California Energy Commission. California State Agencies Collaborative Research Project National Research Council (NRC) Sea Level Rise Study. http://www.energy.ca.gov/2010publications/CAT-1000-2010-005/Research_Collaboration_Case_Studies/Sea_Level_Rise_Study.pdf.

Calvert Global Water Fund. Unparalleled Challenge and Opportunity in Water, May 2010. http://www.calvertgroup.com/NRC/literature/documents/WP10001.pdf?litId=WP10001.

Cape Fear (N.C.) Public Utility Authority (CFPUA). Residential Tips and Information: Home Water Use. http://www.cfpua.org/index.aspx?NID=299.

Chandler, David. A System That's Worth Its Salt. *Massachusetts Institute of Technology News.* March 23, 2010. http://web.mit.edu/newsoffice/2010/desalination-0323.html.

Chang, Kenneth. Like Water off a Beetle's Back. *New York Times.* June 27, 2006. http://www.nytimes.com/2006/06/27/science/27find.html.

Christensen, Lee A. Soil, Nutrient, and Water Management Systems Used in US Corn Production. Electronic Report from the Economic Research Service: Agriculture Information Bulletin No. 774. US Department of Agriculture. April 2002. http://www.ers.usda.gov/publications/aib774/aib774.pdf.

City of Tucson Office of Conservation and Sustainable Development. Business Tools: Green Building and Smart Growth. http://www.tucsonaz.gov/ocsd/business/building/.

Clean Water America Alliance. Managing One Water. *The Clean Water America Alliance's National Dialogue Report.* 2010. http://www.cleanwateramericaalliance.org/pdfs/managing%20one%20water.pdf.

Clean Water America Alliance. What's Water Worth? *The Clean Water America Alliance's National Dialogue Report.* 2010. http://www.cleanwateramericaalliance.org/pdfs/w3report.pdf .

Congressional Budget Office. 2002. *Future Investment in Drinking Water and Wastewater Infrastructure: A CBO Study.* November 2002; page ix. http://www.cbo.gov/doc.cfm?index=3983&type=0.

Cooke, Morris L. *The Future of the Great Plains.* US Government Printing Office, 1936.

Core 77. Sevin Coskun's WASHUP. Core77's Greener Gadgets Design Competition 2008. http://www.core77.com/competitions/GreenerGadgets/projects/4609/.

dePaul, Vincent, Robert Rosman, and Pierre J. Lacombe. Water-Level Conditions in Selected Confined Aquifers of the New Jersey and Delaware Coastal Plain, 2003. New Jersey Department of Environmental Protection. USGS Scientific Investigations Report 2008–5145. http://pubs.usgs.gov/sir/2008/5145/.

Diamond, Jared M. *Collapse: How Societies Choose to Fail or Succeed*. Penguin, 2006.

Díaz, Mireia. The Llobregat Desalination Plant Will Ensure 24% of the Metropolitan Region's Water Consumption. *Council of Barcelona News*. July 20, 2009. http://w3.bcn.es/V01/Serveis/Noticies/V01NoticiesLlistatNoticiesCtl/0,2138,1653_35144087_3_924336390,00.html?accio=detall&home=HomeBCN&nomtipusMCM=Noticia.

Dillon, Sam. Mexico City Journal: Capital's Downfall Caused by Drinking ... of Water. *New York Times*, Jan. 29, 1998. http://www.nytimes.com/1998/01/29/world/mexico-city-journal-capital-s-downfall-caused-by-drinking-of-water.html.

Doremus, Holly D., and A. Dan Tarlock. *Water War in the Klamath Basin: Macho Law, Combat Biology, and Dirty Politics*. Island Press, 2008.

Downey, Dave. 2010. Western Pays Cash to Take Out Grass: 1,500 Murrieta Customers Targeted by Conservation Program. *North County Times–The Californian*. Feb. 23. http://www.nctimes.com/news/local/swcounty/article_c1f881e3-8682-51be-8920-6d88dba35010.html.

Doyle, Patrick, and Natasha Gardner. Dry Times. *5280 Magazine*. April 2010. http://www.5280.com/magazine/2010/04/dry-times.

Economist. A Special Report on Water: For Want of a Drink. May 20, 2010. http://www.economist.com/node/16136302 .

Economist. A Survey of Climate Change: The Heat Is On. Sept. 7, 2006. http://www.economist.com/node/7852924?Story_ID=7852924.

Economist. Cheaper Desalination: Current Thinking. Oct. 29, 2009. http://www.economist.com/node/14743791.

Economist. Monsanto: Lord of the Seeds. Jan. 27, 2005. http://www.economist.com/node/3600040.

Economist. The Great Lakes' Water: Liquid Gold. May 20, 2010: http://www.economist.com/node/16167886.

Eilperin, Juliet. World's Fish Supply Running Out, Researchers Warn. *Washington Post*. http://www.washingtonpost.com/wp-dyn/content/article/2006/11/02/AR2006110200913.html.

Electrolux Design Lab. http://www.electroluxdesignlab.com/.

Electrolux Group. http://group.electrolux.com/en/.

Environmental Business International (EBI). 2010. Water and Wastewater. *Environmental Business Journal* XXIII: 11: 2.

Environmental Impacts (Purdue University). Environmental Impacts of Home Lawn Care. http://www.purdue.edu/envirosoft/lawn/src/environmental.htm.

Ercin, A., M. Ertug, Martinez Aldaya, and Arjen J. Hoekstra. Corporate Water Footprint Accounting and Impact Assessment: The Case of the Water Footprint of a Sugar-Containing Carbonated Beverage, *Water Resource Management*, Oct. 23 2010.

Fatta-Kassinos, Despo, Kai Bester, and Klaus Kümmerer, eds. *Xenobiotics in the Urban Water Cycle*. Springer, 2010.

Flores, Heather C. *Food Not Lawns: How to Turn Your Yard into a Garden and Your Neighborhood into a Community*. Chelsea Green Publishing, 2006.

Franklin D. Roosevelt's Dedication Day Speech, Sept. 30, 1935. Available at University of Virginia American Studies Program and the 1930s Project. http://xroads.virginia.edu/~MA98/haven/hoover/fdr.html.

Fuller, Thomas. Countries Blame China, Not Nature, for Water Shortage. *New York Times,* April 1, 2010. http://www.nytimes.com/2010/04/02/world/asia/02drought.html.

FutureFarmIndustriesCRC.SaltTolerantCereal(FP12).FutureFarmOnline.2009. http://www.futurefarmonline.com.au/research/future-cropping-systems/salt-tolerant-cereal.htm.

Future Farm Online. http://www.futurefarmonline.com.au/.

Gasson, Christopher. A New Virtual Reality. *Global Water Intelligence* 9:9. September 2008. http://www.globalwaterintel.com/archive/9/9/analysis/a-new-virtual-reality.html.

Gibbison, Godfrey A., and James Randall. The Salt Water Intrusion Problem and Water Conservation Practices in Southeast Georgia, USA. *Water and Environment Journal* 20, no. 4 (12, 2006): 271–281.

Glantz, Michael H. *Drought Follows the Plow: Cultivating Marginal Areas*. Cambridge University Press, 1994.

Gleick, Peter H. *Bottled and Sold: The Story Behind Our Obsession with Bottled Water*. Island Press, 2010.

Gleick, Peter H., Heather Cooley, and David Katz. *The World's Water, 2006–2007: The Biennial Report on Freshwater Resources*. Island Press, 2006.

Gleick, Peter H., Heather Cooley, and Mari Morikawa. *The World's Water 2008–2009: The Biennial Report on Freshwater Resources*. Island Press, 2008.

Glennon, Robert Jerome. *Unquenchable: America's Water Crisis and What to Do About It*. Island Press, 2009.

Global Water Intelligence. 2007. Arizona Water Rights Auction Tops $20/m^3. *Global Water Intelligence* 8:11. November 2007, page 22. http://www.globalwaterintel.com/archive/8/11/general/arizona-water-rights-auction-tops-20msup3sup.html.

Global Water Intelligence. 2010a. Transforming the World of Water: The Book of the Global Water Summit 2010, Paris. http://files.globalwaterintel.com/dl/20101018_Transforming_the_world_of_water.pdf.

Global Water Intelligence. 2010b. Global Water Intelligence Water Tariff Survey, September, page 31.

Gordon, Greg. *Landscape of Desire: Identity and Nature in Utah's Canyon Country*. Utah State University Press, 2003.

Greater Milwaukee Committee. The Milwaukee 7 Water Council. http://www.gmconline.org/index.php?option=com_content&task=view&id=150&Itemid=76

Great Lakes Commission. Great Lakes Basin Compact. http://www.glc.org/about/glbc.html.

Great Lakes Diversion and Water Use. *Stop Asian Carp: Protect Our Great Lakes. A Project of Michigan Attorney General Mike Cox.* at http://www.stopasiancarp.com/agcox.html.

Halverson, Anders. *An Entirely Synthetic Fish: How Rainbow Trout Beguiled America and Overran the World.* Yale University Press, 2010.

Hart, Andrea. 2009. US Water Use Declines, But Points to Troubling Trends, Says USGS Report. *Circle of Blue,* Oct. 29. http://www.circleofblue.org/waternews/2009/world/u-s-water-use-declined-from-2000-to-2005-part-of-growing-trend-usgs-report-finds/.

Hoekstra, A.Y., and A.K. Chapagain. 2008. *Globalization of Water: Sharing the Planet's Freshwater Resources.* Blackwell Publishing:Oxford, UK.

Hoekstra, Arjen. A Comprehensive Introduction to Water Footprints. 2011. http://www.waterfootprint.org/downloads/WaterFootprint-Presentation-General.pdf.

Hoffmann, Stephen J. *Planet Water: Investing in the World's Most Valuable Resource.* John Wiley and Sons, 2009.

Humayun, Faisal. 2010. Blue Gold: An Excellent Investment Opportunity. *Alrroya.com.* April 7. http://english.alrroya.com/content/blue-gold-excellent-investment-opportunity.

Humes, Edward. *Eco Barons: The Dreamers, Schemers, and Millionaires who Are Saving Our Planet.* Ecco, 2009.

Hutson, Susan S., Nancy L. Barber, Joan F. Kenny, Kristin S. Linsey, Deborah S. Lumia, and Molly A. Maupin. *Estimated Use of Water in the United States in 2000.* US Geological Survey Circular 1268. http://pubs.usgs.gov/circ/2004/circ1268/.

Hydrological Society of South Australia, History Trust of South Australia, Stormwater Industry Association, Institution of Engineers, and Australia, South Australian Division. Engineering Heritage Branch. *Water History: Lessons for the Future, Friday 28 September 2001, PS Murray Queen, Goolwa, Australia.* Hydrological Society of South Australia, 2001.

Intel. Water Footprint Analysis. *Intel 2009 Corporate Responsibility Report.* http://www.intel.com/Assets/PDF/Policy/CSR-2009.pdf#page=42.

International Commission on Large Dams (ICOLD). 1998. ICOLD World Register of Dams, Computer Database, Paris. In *Dams and Development: The Report of the World Commission on Dams*, November 2001, 1:8.

International Organization for Dew Utilization. http://www.opur.fr/index.htm.

International Paper. *Global Water Use.* 2010. http://www.internationalpaper. com/us/en/company/Sustainability/PF_GeneralContent_1_3601_3601.html.

ITT News. ITT's Value of Water Survey Reveals That Americans Are Ready to Fix Our Nation's Crumbling Water Infrastructure. Oct. 27, 2010. http://www. itt.com/news/press-releases/release_20101027.asp .

Jenkins, Virginia Scott. *The Lawn: A History of an American Obsession.* Smithsonian Institution Press, 1994.

Jha, Alok. 2008. Seawater Greenhouses to Bring Life to the Desert. *The Guardian.* Sept. 2. http://www.guardian.co.uk/environment/2008/sep/02/alternativeenergy.solarpower#.

Jonsson, Patrik. Drought-Stricken Georgia, Eyeing Tennessee River, Revives Old Border Feud. *Christian Science Monitor.* Feb. 15, 2008. http://www.csmonitor. com/USA/2008/0215/p02s02-usgn.html.

JP Morgan Securities, Piet Klop, and Fred Wellington. 2008. *Watching Water: A Guide to Evaluating Corporate Risks in a Thirsty World.* World Resources Institute. March 31. http://www.wri.org/publication/watching-water.

Kameri-Mbote, Patricia. Water, Conflict, and Cooperation: Lessons from the Nile River Basin. *Nile Media Network.* June 25, 2010. http://www.nilemedia network.com/index.php?option=com_content&view=article&id=70:water-conflict-and-cooperation-lessons-from-the-nile-river-basin&catid=37:articles.

Kealey, Lucy. Progress Continues for Tolerant Cereal. Future Farm Industries. http:// www.futurefarmonline.com.au/_literature_58684/Progress_continues_for _tolerant_cereal.

Klein, Bradley S., and *Golfweek* magazine. *A Walk in the Park: Golfweek's Guide to America's Best Classic and Modern Golf Courses.* Sports Publishing LLC, 2004.

Knechtel, John. *Water.* MIT Press, 2009.

Kostigen, Thomas M. *The Green Blue Book: The Simple Water-Savings Guide to Everything in Your Life.* Rodale Books, 2010.

Kurtenbach, Elaine. As US Debates Projects, China Builds Them. *Courier Post Online,* Dec. 26, 2010. http://www.courierpostonline.com/article/20101226/ BUSINESS/12260323/As-U-S-debates-projects-China-builds-them.

Laden, Karl. *Antiperspirants and Deodorants.* CRC Press, 1999.

LaMonica, Martin. Wave-Powered Desalination Pump Permitted in Gulf. Green Tech. *CNET News.* May 28, 2010. http://news.cnet.com/8301-11128_3-20006269-54.html.

Lavelle, Marianne. Money and Business: A National Water Crisis Is on the Verge of Gushing. *US News and World Report.* May 27, 2007. http://www.usnews. com/usnews/biztech/articles/070527/4water.htm.

Leake, S.A. Land Subsidence from Ground-Water Pumping. Workshop on Impact of Climate Change and Land Use in the Southwestern United States, July 1997. US Department of the Interior, US Geological Survey. http://geo-change.er.usgs.gov/sw/changes/anthropogenic/subside/.

Lee, Katie, and Terry Tempest Williams. *All My Rivers Are Gone: A Journey of Discovery Through Glen Canyon.* Big Earth Publishing, 1998.

Lenzer, Anna. Fiji Water: Spin the Bottle. *Mother Jones.* Sept.-Oct. 2009. http://motherjones.com/politics/2009/09/fiji-spin-bottle .

Leslie, Jacques. *Deep Water: The Epic Struggle over Dams, Displaced People, and the Environment.* Macmillan, 2006.

Lindsey, Rebecca. Looking for Lawns. *NASA Earth Observatory.* Nov. 8, 2005. http://earthobservatory.nasa.gov/Features/Lawn/.

Macan-Markar, Marwaan. China Flexes Hydropower Muscle. *Daily Mirror* (Sri Lanka), Aug. 30, 2010. http://print.dailymirror.lk/opinion1/19921.html.

Mann, Charles. The Rise of Big Water. *Vanity Fair.* May 2007. http://www.vanity fair.com/politics/features/2007/05/bigwater200705.

Margolin, Victor, and Richard Buchanan. *The Idea of Design.* MIT Press, 1995.

Marks, Susan J. *Aqua Shock: The Water Crisis in America.* Bloomberg Press, 2009.

Marsden, William. *Stupid to the Last Drop: How Alberta Is Bringing Environmental Armageddon to Canada (and Doesn't Seem to Care).* Random House, Inc., 2007.

Martin, Andrew. Tap Water's Popularity Forces Pepsi to Cut Jobs. *New York Times,* Oct. 14, 2008. http://www.nytimes.com/2008/10/15/business/15pepsi.html?_r=1&ref=indraknooyi.

Mauser, Wolfram. *Water Resources: Efficient, Sustainable and Equitable Use.* Haus, 2009.

Maxwell, Steve, ed. *The Business of Water: A Concise Overview of Challenges and Opportunities in the Water Market.* American Water Works Association, 2008.

McCully, Patrick. *Silenced Rivers: The Ecology and Politics of Large Dams.* Zed Books, 2001.

McTigue, Nancy E., and James M. Symons. *The Water Dictionary.* 2nd ed. AWWA, 2010.

Means, Edward G., Awwa Research Foundation, Lorena Ospina, American Water Works Association, Nicole West, and IWA Publishing. *A Strategic Assessment of the Future of Water Utilities.* American Water Works Association, 2006.

Means, Edward G., Awwa Research Foundation, US Environmental Protection Agency. *Communicating the Value of Water: An Introductory Guide for Water Utilities.* Awwa Research Foundation, 2008.

Miller, Barbara. Breakthrough Could Lead to Drought-Resistant Plants. *Australian Broadcasting Corporation News,* Feb. 28, 2008. http://www.abc.net.au/news/stories/2008/02/28/2175160.htm.

Miller, Lisa. Bless This Bottled Water. *Newsweek,* Dec. 8, 2007. http://www.newsweek.com/2007/12/08/bless-this-bottled-water.html.

Milwaukee Water Council. http://www.milwaukee7-watercouncil.com/wiki/show/Main.

Morrison, Jason I., Mari Morikawa, Michael Murphy, and Peter Schulte. *Water Scarcity and Climate Change: Growing Risks for Businesses and Investors.* Ceres, 2009. http://www.ceres.org/Document.Doc?id=406.

Morse, Dan, and Katherine Shaver. Water Main Break Forces Dramatic Rescue of Nine. *Washington Post.* Dec. 24, 2008. http://www.washingtonpost.com/wp-dyn/content/article/2008/12/23/AR2008122302853.html.

Moss, Richard J. *Golf and the American Country Club.* University of Illinois Press, 2001.

Mullin, Megan. *Governing the Tap: Special District Governance and the New Local Politics of Water.* MIT Press, 2009.

National Club Association Press Releases. http://www.natlclub.org/club/scripts/view/view_insert.asp?NS=PUBLIC&INAME=PressReleases.

National Club Association: National Golf Day 2010. http://www.nationalclub.org/club/scripts/library/view_document.asp?CLNK=1&GRP=&NS=PUBLIC&DID=98547&APP=80.

Newell, Frederick H. *Water Resources: Present and Future Uses.* New Haven: Yale University Press, 1920.

Newmont Mining Corporation. For Mining Firms, It All Comes Down to H2O. Adapted from *Waste & Recycling News.* http://www.newmont.com/features/our-environment-features/For-Mining-Firms-It-All-Comes-Down-to-H2O%20.

Newmont Mining Corporation. Shaping Sustainable Development in Peru. *Beyond the Mine: the Journey Towards Sustainability.* http://www.beyondthemine.com/2009/?l=2&pid=240&parent=253&id=428.

Niman, Nicolette Hahn. *Righteous Porkchop: Finding a Life and Good Food Beyond Factory Farms.* William Morrow, 2009.

Ogilvie, Felicity. Tasmanian Government Rejects Pipeline Plan. *PM* (Australian Broadcasting Corporation). Feb. 4, 2009. http://www.abc.net.au/pm/content/2008/s2482512.htm.

Ohio Department of Natural Resources. Federal Statute on Great Lakes Water Diversions. http://ohiodnr.com/water/planing/greatlksgov/fedstatut/tabid/4053/Default.aspx.

Oleson, John Peter. *Greek and Roman Mechanical Water-Lifting Devices: The History of a Technology.* Springer, 1984.

Ozler, Levent. Electrolux Rockpool: Waterless Dishwasher. *Dexigner.* Nov. 23, 2004. http://www.dexigner.com/news/3239.

Pacific Institute. Peter Gleick Discusses "Peak Water," China's Water Crisis, Climate Change Impacts. Pacific Institute Press Release, Jan. 5, 2009. http://www.pacinst.org/press_center/press_releases/woodrow_wilson_2008_2009.html.

Pacific Institute. Report Warns Businesses and Investors About Growing Water Scarcity Impacts from Climate Change. Feb. 26, 2009. http://www.ceres.org/Page.aspx?pid=1041.

Pacific Institute. US Per Capita Water Use Falls to 1950s Levels: Analysis of USGS Data Shows That Efficiency Is Effective, Demand Is Not Endless. March 11, 2004. http://www.pacinst.org/press_center/usgs/.

Parish, Billy. Judge Withdraws Black Mesa Mining Permit. *Native American Times.* Jan. 8, 2010. http://nativetimes.com/index.php?option=com _content&view=article&id=2842:judge-withdraws-black-mesa-mining-permit&catid=56&Itemid=32.

Pearce, Fred. *When the Rivers Run Dry: Water, the Defining Crisis of the Twenty-First Century.* Beacon Press, 2007.

Peet, John. A Survey of Water: Priceless. *Economist.* July 17, 2003. http://www.economist.com/node/1906846.

Pilz, David. Charting the Colorado Plateau Revisited. Sustainable Development Workshop, Economics Department, Colorado College. http://www.colorado college.edu/Dept/EC/Faculty/Hecox/CPWebpage/issuespageTurner.htm.

Pinsent Masons. 2006. *Pinsent Masons Water Yearbook 2006–2007.* London. http://www.nedwater.eu/documents/PinsentMasonsWaterYearbook2006–2007.pdf.

Pinsent Masons. 2009. *Pinsent Masons Water Yearbook 2009–2010.* London. http://wateryearbook.pinsentmasons.com/PDF/Water%20Yearbook%20 2009-2010.pdf.

Pollan, Michael. *Second Nature: A Gardener's Education.* Grove Press, 2003.

Pollan, Michael. *The Omnivore's Dilemma: A Natural History of Four Meals.* Penguin Press, 2006.

Porter, Eliot, and David Ross Brower. *The Place No One Knew: Glen Canyon on the Colorado.* Peregrine Smith Books, 1988.

Postel, Sandra. *Last Oasis: Facing Water Scarcity.* W. W. Norton & Company, 1997.

Postel, Sandra. *Pillar of Sand: Can the Irrigation Miracle Last?* New York: W. W. Norton & Company, 1999.

Powell, James Lawrence. *Dead Pool: Lake Powell, Global Warming, and the Future of Water in the West.* University of California Press, 2008.

Primeau, Liz. *Front Yard Gardens: Growing More Than Grass.* Second Edition, Updated and Expanded. Firefly Books, 2010.

Qing, Dai, John Thibodeau, and Philip B. Williams. *The River Dragon Has Come!: The Three Gorges Dam and the Fate of China's Yangtze River and Its People.* M.E. Sharpe, 1998.

Reisner, Marc. *Cadillac Desert: The American West and Its Disappearing Water.* Penguin, 1993.

Renewable Energy Development. 2008. Solar Energy: Mojave Solar Park (CSP). March 19. http://renewableenergydev.com/red/solar-energy-mojave-solar-park-csp/. See also http://www.esolar.com/.

Report. California. Agricultural Experiment Station. Berkeley, 1922.

Rodwan, John G. Jr. Bottled Water 2009. Challenging Circumstances Persist: Future Growth Anticipated. International Bottled Water Association. http://www.bottledwater.org/files/2009BWstats.pdf.

Rogers, Peter. Facing the Freshwater Crisis. *Scientific American.* July 23, 2008. http://www.scientificamerican.com/article.cfm?id=facing-the-freshwater-crisis.

Rothfeder, Jeffrey. *Every Drop for Sale: Our Desperate Battle over Water in a World About to Run Out.* Penguin, 2004.

Royte, Elizabeth. A Tall, Cool Drink of ... Sewage? *New York Times.* Aug. 10, 2008. http://www.nytimes.com/2008/08/10/magazine/10wastewater-t.html.

Royte, Elizabeth. *Bottlemania: How Water Went on Sale and Why We Bought It.* Bloomsbury, 2008.

Salina, Irena. *Written in Water: Messages of Hope for Earth's Most Precious Resource.* National Geographic Books, 2010.

Schneider, Keith. 2010. In Era of Climate Change and Water Scarcity, Meeting National Energy Demand Confronts Major Impediments. *Circle of Blue.* Sept. 22. http://www.circleofblue.org/waternews/2010/world/in-era-of-climate-change-and-water-scarcity-meeting-national-energy-demand-confronts-major-impediments/.

School of Freshwater Sciences, University of Wisconsin–Milwaukee. Great Lakes Water Institute. http://www.glwi.uwm.edu/.

Scott, Frank J. *The Art of Beautifying Suburban Home Grounds of Small Extent: The Advantages of Suburban Homes Over City or Country Homes; The Comfort and Economy of ... Trees and Shrubs Grown in the United States.* D. Appleton & Co., 1870.

Segerfeldt, Fredrik. *Water for Sale: How Business and the Market Can Resolve the World's Water Crisis.* Cato Institute, 2005.

Shiva, Vandana. *Water Wars: Privatization, Pollution and Profit.* South End Press, 2002.

Smith, Gary C. The Future of the Beef Industry. *Proc., The Range Beef Cow Symposium XIX, December 6–8, 2005, Rapid City, S.D.* http://beef.unl.edu/beefreports/symp-2005-02-XIX.pdf.

Smith, Mark, Dolf De Groot, and Ger Bergkamp. *Pay: Establishing Payments for Watershed Services.* IUCN, 2006.

Smith, Morgan. Lawsuit Could Determine Future of Groundwater. *The Texas Tribune.* April 22, 2010. http://www.texastribune.org/texas-environmental-news/water-supply/lawsuit-could-determine-future-of-groundwater/.

Solomon, Steven. *Water: The Epic Struggle for Wealth, Power, and Civilization.* HarperCollins, 2010.

Specter, Michael. The Last Drop: Confronting the Possibility of a Global Catastrophe. *The New Yorker* Oct. 23, 2006. http://www.newyorker.com/archive/2006/10/23/061023fa_fact1.

Steinberg, Ted. *American Green: The Obsessive Quest for the Perfect Lawn*. Norton, 2006.

Steinberg, Ted. Lawn Mores. *Los Angeles Times*. March 18, 2006. http://articles. latimes.com/2006/mar/18/opinion/oe-steinberg18.

Stilgoe, John R. *Shallow Water Dictionary: A Grounding in Estuary English*. Princeton Architectural Press, 2003.

Sugano, Shigeo S., Tomoo Shimada, Yu Imai, Katsuya Okawa, Atsushi Tamai, Masashi Mori, Ikuko Hara-Nishimura. Stomagen Positively Regulates Stomatal Density in *Arabidopsis*. *Nature*. 463, 241–244. Jan. 14 2010. http://www. nature.com/nature/journal/v463/n7278/full/nature08682.html.

SummitCaseWaterEquityInvesting2010.pdf. http://www.summitglobal.com/doc uments/SummitCaseWaterEquityInvesting2010.pdf.

SustainAbility. Issue Brief: Water. http://www.sustainability.com/library/issue-brief-water.

Thomas, William L. *Man's Role in Changing the Face of the Earth*. University of Chicago Press, 1956.

United Nations. 2010. *Factsheet on Water and Sanitation*. International Decade for Action: Water for Life, 2005-2015. http://www.un.org/waterforlifedecade/ factsheet.html.

US Congressional Research Service. Energy Independence and Security Act of 2007: A Summary of Major Provisions. http://energy.senate.gov/public/_files/ RL342941.pdf.

US Department of Agriculture (USDA). Selected Crops Irrigated and Harvested by Primary Method of Water Distribution, United States: 2008. Farm and Ranch Irrigation Survey. 2007 Census of Agriculture. http://www.agcensus.usda.gov/ Publications/2007/Online_Highlights/Farm_and_Ranch_Irrigation_Sur vey/fris08_1_30.pdf.

USDA and National Agricultural Statistics Service. Census of Agriculture. http:// www.agcensus.usda.gov/.

US Environmental Protection Agency (USEPA). 1998. President Clinton: Clean, Safe Water for All Americans. *Clean Water Action Plan*. EPA press release, February 19. http://www.epa.gov/history/topics/cwa/03.htm.

USEPA. 2001. Contaminants Regulated Under the Safe Drinking Water Act. www.epa.gov/safewater/contaminants/pdfs/contam_timeline.pdf.

USEPA. 2009a. Drinking Water Infrastructure Needs Assessment and Survey, Fourth Report to Congress, February 2009, 9.

USEPA. 2009b. Water on Tap: What You Need to Know. http://water.epa.gov/ drink/guide/.

USEPA and US Department of Energy. *ENERGY STAR Qualified Clothes Washers*. http://www.energystar.gov/ia/products/appliances/clotheswash/Qualified_ CW_2010_Criteria.xls.

US Geological Survey (USGS). Groundwater Information: *Land Subsidence in the United States.* (USGS Fact Sheet 165-00). http://water.usgs.gov/ogw/pubs/fs00165/.

US Government Accountability Office. Energy and Water: Preliminary Observations on the Links Between Water and Biofuels and Electricity Production. http://www.gao.gov/products/GAO-09-862T.

USGS. *Land Subsidence in the United States.* USGS Circular 1182, September 2005. http://pubs.usgs.gov/circ/circ1182/.

USGS. Water Use in the United States. http://water.usgs.gov/watuse/.

Valuing Ground Water: Economic Concepts and Approaches. National Academy Press, 1997.

Villiers, Marq De. *Water: The Fate of Our Most Precious Resource.* Houghton Mifflin Harcourt, 2001.

Walton, Brett. 2010. The Rising Cost of Settling the American Desert. *Circle of Blue,* Oct. 11. http://www.circleofblue.org/waternews/2010/world/power-plant-that-moves-torrent-of-water-uphill-considers-closing/.

Washington Suburban Sanitary Commission. 2008 Continues Water Main Break, Leak Trend. Jan. 6, 2009. http://www.wsscwater.com/home/jsp/misc/genericNews.faces?pgurl=/Communication/NewsRelease/2009/2009-01-06.html.

Water Facts and Trends. World Business Council for Sustainable Development. http://www.wbcsd.org/plugins/DocSearch/details.asp?type=DocDet&ObjectId=MTYyNDk.

Water Footprint Network. Product Gallery: Beef. http://www.waterfootprint.org/?page=files/productgallery&product=beef.

Watts, Jonathan. 2009. China at the Crossroads: Thirst of the Cities Drives the Giant Drills to Water China's Parched North. *The Guardian.* May 18. http://www.guardian.co.uk/world/2009/may/18/china-water-crisis

Waughray, Dominic. *Water Security: Managing at the Water-Food-Energy-Climate Nexus: The World Economic Forum Water Initiative.* Island Press, 2011.

Webber, Michael E. Catch-22: Water vs. Energy. *Scientific American Earth 3.0.* http://www.ce.utexas.edu/prof/mckinney/ce397/Topics/Sustainability/Webber2008.pdf.

Webber, Michael E. Energy Versus Water: Solving Both Crises Together. *Scientific American.* Oct. 22, 2008. http://www.scientificamerican.com/article.cfm?id=the-future-of-fuel.

Wild, Daniel, Marc-Olivier Buffle, and Junwei Hafner-Cai. *Water—A Market of the Future.* SAM Study. 2010. http://www.sam-group.com/downloads/studies/waterstudy_e.pdf.

Will, George F. *Bunts.* Simon and Schuster, 1999.

Wood, Chris. *Dry Spring: The Coming Water Crisis of North America.* Raincoast Books, 2008.

Workman, James G. *Heart of Dryness: How the Last Bushmen Can Help Us Endure the Coming Age of Permanent Drought.* Walker & Co., 2009.

World Bank. 2010. 40 Percent of World Population Do Not Use Improved Sanitation Facilities. July 19. http://data.worldbank.org/news/40prct-wrld-pop-dont-use-imprvd-sanitation.

Yates, Scott. Florida Utility Tests Stormwater Retrieval. *AWWA Streamlines* 1:26. Dec. 22, 2009. http://www.awwa.org/Publications/StreamlinesArticle.cfm?ItemNumber=52619.

INDEX

Note: *f.* indicates figure; *t.* indicates table.